RENEWABLE
NATURAL
RESOURCES

ECONOMIC
INCENTIVES
FOR IMPROVED MANAGEMENT

ORGANISATION FOR ECONOMIC CO-OPERATION AND DEVELOPMENT

Pursuant to article 1 of the Convention signed in Paris on 14th December 1960, and which came into force on 30th September 1961, the Organisation for Economic Co-operation and Development (OECD) shall promote policies designed:

- to achieve the highest sustainable economic growth and employment and a rising standard of living in Member countries, while maintaining financial stability, and thus to contribute to the development of the world economy;
- to contribute to sound economic expansion in Member as well as non-member countries in the process of economic development; and
- to contribute to the expansion of world trade on a multilateral, non-discriminatory basis in accordance with international obligations.

The original Member countries of the OECD are Austria, Belgium, Canada, Denmark, France, the Federal Republic of Germany, Greece, Iceland, Ireland, Italy, Luxembourg, the Netherlands, Norway, Portugal, Spain, Sweden, Switzerland, Turkey, the United Kingdom and the United States. The following countries acceded subsequently through accession at the dates indicated hereafter: Japan (28th April 1964), Finland (28th January 1969), Australia (7th June 1971) and New Zealand (29th May 1973).

The Socialist Federal Republic of Yugoslavia takes part in some of the work of the OECD (agreement of 28th October 1961).

Publié en français sous le titre:

**RESSOURCES NATURELLES
RENOUVELABLES**
Incitations économiques
pour une meilleure gestion

This report has been prepared as part of the programme of the OECD's Environment Committee on Natural Resource Management.

The report itself is the product of a workshop held at the OECD dealing with the role of economic incentives for improved management of water, forests and land. It contains the papers submitted for the workshop and the conclusions arising out of them.

Also available

PRICING OF WATER SERVICES (1987)
(97 87 02 1) ISBN 92-64-12921-9 146 pages   £8.00   US$17.00   FF80.00 DM38.00

OECD ENVIRONMENTAL DATA/DONNÉES OCDE SUR L'ENVIRONNEMENT – COMPENDIUM 1987. Bilingual
(97 87 05 3) ISBN 92-64-02960-5 366 pages   £20.00   US$42.00   FF200.00 DM86.00

WATER POLLUTION BY FERTILIZERS AND PESTICIDES (1986)
(97 86 02 1) ISBN 92-64-12856-5 144 pages   £6.00   US$12.00   FF60.00 DM27.00

RURAL PUBLIC MANAGEMENT (1986)
(42 86 02 1) ISBN 92-64-12858-1 86 pages   £5.00   US$10.00   FF50.00 DM25.00

Prices charged at the OECD Bookshop.

*THE OECD CATALOGUE OF PUBLICATIONS and supplements will be sent free of charge
on request addressed either to OECD Publications Service,
2, rue André-Pascal, 75775 PARIS CEDEX 16, or to the OECD Distributor in your country.*

TABLE OF CONTENTS

Management of natural resources have been in the forefront of the environmental debate in recent years for a number of reasons. Firstly, it has come to be accepted that sustainable economic growth compatible with long run environmental objectives, including conservation of natural resources for future generations, depends on efficient management of natural resources. Secondly, resource management policies are recognised as fertile ground for the development of anticipatory environmental policies, including pollution control. Thirdly, resource management policies are looked upon to provide and to safeguard the natural heritage, including the recreational environment in protected natural areas.

In April 1987, the OECD's Environment Directorate organised, as part of its on-going programme of work on natural resources, a workshop of natural resources experts. The purpose of the workshop was:

a) to examine the main factors giving rise to economic and environmental mismanagement of renewable natural resources, with particular reference to government intervention in the form of 'incentives';

b) to suggest policy measures to remove inefficiencies of resource mismanagement that lead to 'government failures' and 'market failures'.

The participating experts came from government, private enterprises and universities. The workshop was organised around main presentations for each of the three resource categories (water, forests, land) with papers by the other participants commenting on or adding to the discussion on the main papers. This publication presents the workshop papers as revised to reflect additional issues raised at the workshop.

Governments in all OECD Member countries are heavily involved in the management of natural resources. Such resources are often state-owned, and where privately owned, their management is strongly influenced by government policies. Current resource management policies range from redefining and reallocating property rights to water, to zoning land for specific uses, or to encouraging market forces to play a lesser or greater role.

Using concrete examples, the expert submissions discussed the causes and potential remedies for 'policy failures' typical of current approaches to natural resources management in a variety of countries. These were characterised as 'government failures' attributable to public ownership and management of natural resources and the broad range of government incentives aimed at private owners, or 'market failures' derived from private ownership and exploitation. Flowing from this dichotomy, further discussion centered on the difficulties of defining and ascribing private property rights to a resource in such a way that private sector managers, despite being more immediately concerned with good (i.e. profitable) economic exploitation, take broader developmental goals into account.

A variety of recent concepts in environmental economics were discussed, including social costs of resource use, total economic value, options and existence values, natural resource accounting, environmentally adjusted national accounts, water release and capacity sharing, resource rents, below cost sales, negotiated and consensus management.

The participants suggested a variety of possible policy options that could help political decision-makers and natural resource managers to select the appropriate mix of market mechanisms, economic incentives and government intervention and regulation to achieve the socially optimal use of renewable natural resources for present and future generations.

Finally, the workshop identified a number of issues which might be appropriately focused on in further joint efforts by OECD Member countries, including:

-- further examination of the concept of sustainability as regards natural resources;

-- development of methodologies for measuring natural resource damage;

-- surveying the influence of benefit estimation on decision-making;

-- further exploration of various systems of natural resource and environmental accounting;

-- assessment of the extent and importance of natural resource mismanagement;

-- comparative analyses of differing regimes of public lands' management;

-- reformulation of development assistance to take into account the importance of the environment to the development process.

Both the views expressed in the papers and the conclusions represent the personal opinions of the invited experts. Nevertheless, they can be regarded as reflecting widely-held views in a broad cross-section of OECD Member countries.

Chapter 1

ECONOMIC INCENTIVES AND RENEWABLE
NATURAL RESOURCE MANAGEMENT

Theoretical Context and Preview of the Issues

Prof. David W. Pearce

I. INTRODUCTION: NATURAL RESOURCES AND THE DEVELOPMENT PROCESS

Although possible absolute global scarcity of natural resources was discussed widely in the late 1960s and early 1970s, and often in alarmist fashion, few authorities now express concern about such scarcity in the 1980s. Global concerns exist, particularly over the future of receiving environments such as the atmosphere, stratosphere and the oceans. But there is now markedly less talk about 'running out' of energy or minerals. The early concerns are however still relevant, if not more so, in respect of the absolute scarcity and the quality of natural resources in developing economies.

For OECD economies, problems of quality also exist, particularly in respect of the state of some forest resources, water, wildlife and human recreational habitat. Natural resource use is also inseparable from problems of pollution. Thermodynamic principles necessarily imply that resource use rates affect the flows of waste and residuals to the natural environment. The complex of ecological linkages in any economy further dictates that resource use in one sector of the economy will frequently appear as a residuals problem in another sector of the economy. Fertilizer application to land affects water quality in households, fossil fuel combustion in power stations affects groundwater quality often many miles distant, and so on.

Nor can OECD concerns be divorced from developing economy resource problems since:

a) OECD imports of natural resources generate foreign exchange for developing countries, whilst simultaneously providing the source of demand for activities which partly contribute to resource degradation;

b) OECD aid and lending policy is increasingly focussing on renewable resource augmentation in developing countries;

11

c)   OECD and developing economies are interlinked through trade and a common concern for the future well-being of all populations; and

d)   resource degradation in developing economies imposes costs on OECD countries through, for example, loss of valued habitats and species.

Leaving aside the developing -- developed economy linkages for the moment, the issue that arises is whether the natural resource base is being efficiently managed.   If not, then there are costs in terms of foregone development potential.   Inefficient resource use simply uses up more resources than need be to produce a given unit of economic output or 'welfare'. Inefficient use thus depletes an exhaustible resource more quickly than necessary, or incurs unwarranted costs of regenerating a renewable resource. Inefficient resource use also generates more demands on receiving assimilative environments and hence lowers their quality, perhaps to the point of incurring real risks of irreversible damage.

In so far as all these costs are shifted forward in time, they are borne either by the current generation in future years or by future generations themselves.   While the 'time horizon' of governments is often short, as dictated by political considerations, that of international organisations such as OECD incorporates the long term.   Morality demands that the long-term prospects for development be part of the rational use of natural resources.

Finally, the development process consists of much more than sustained increases in real consumption levels, important though these are.   It must also incorporate social objectives such as education, physical and mental health and spiritual values.   In an increasingly stressful world, proximity to natural environments and the chance to enjoy them becomes a major factor in improving the quality of life.   A convenient way of summarising this objective is to refer to development as the objective of society.   Development subsumes economic growth in the conventional sense of increases in GNP, but is wider in that it encompasses distributional and welfare considerations.

Overall, then, the background theme to the analysis of economic incentives and natural resources is that the socially optimal use of natural resources is an integral part of the process of sustained economic development.

II. THE APRIL 1987 PARIS WORKSHOP

Taking this focus of the relationship between natural resources and development as its backdrop, OECD assembled a body of experts in April 1987 in Paris for an initial Workshop designed to elucidate the issues that arise in the context of devising incentives for improved natural resource management. In turn, the issues will be related to a Work Programme for the OECD Environment Division for 1988 -- 1990 in the first instance.   In many respects this dimension of OECD's work is a mirror image of the substantial programme of activity in the 1970s and 1980s on pollution.   Many of the questions asked about pollution can and should be asked about natural resource use.   These are:

-- How might the extent of 'damage' done by the misuse of natural resources be measured?

-- What are the chief factors giving rise to resource mismanagement?

-- What are the policy measures available to Member countries for the correction of resource mismanagement?

The focus of the Workshop was on the second and third questions. There was a presumption that the techniques developed for measuring pollution damage would, by and large, be extendable to measuring resource misuse damage, although that presumption needs testing.

'Causal' analysis and policy measures are, of course, closely interrelated. While the invited papers generally discussed the causal factors within the context of two broad classifications -- market failure and government failure -- some contributors felt that the distinction was unhelpful, particularly those who felt that resource economists still had not appreciated what they regarded as a revolution in paradigmatic thinking arising from the property rights school. By and large, this school argues that most resource problems can be solved by a proper and rigorous definition of property rights in the resource. Issues such as 'proper' pricing remain significant but take a distinct second place to property rights classifications. The contribution by Paterson (Chapter 3) is the most forceful statement to this effect presented to the Workshop, and it repays careful study.

The term 'incentives' can be interpreted widely to include, for example, giving greater emphasis to market forces which themselves provide the incentives to better resource management. It may mean governments intervening to zone areas for exclusive use and deliberately preventing use conflicts that arise from market forces. It may also mean redefining property rights with a government agency acting as 'manager' of the resource but with competing users having property rights to quantities of the resource in situ (as Paterson prefers -- see Chapter 3).

III. THE NATURAL RESOURCES IN QUESTION

The focus of concern in the OECD programme of work on natural resources is land-based renewable resources. The neglect of exhaustible resources is justified by the fact that most expressions of concern within OECD countries tend to be about renewable resources: forests, soil (a mix of renewable and exhaustible components) and water. As Frederick (Chapter 2) notes in his opening remarks, water in particular promises to be the next scarce resource to cause growing alarm. Other contributors stressed the importance of habitats, resources that may well be considered to be non-renewable in that regeneration could not be brought about, at least in the form of replicating a lost area. The whole issue of "amenity land" is critical, especially in the more crowded nations of Europe. The neglect of non land-based renewable resources, primarily coastal waters and oceans, is occasioned by the need to begin the OECD work programme modestly. It would seem essential, however, at an early stage in OECD's further work in this area, to refine the concept of sustainable development.

IV. TWO ORGANISING PRINCIPLES FOR ANALYSING RESOURCE MANAGEMENT

Analysis of social importance of resource misuse is facilitated by two related concepts. The first is the social opportunity cost of resource use. This relates to the optimal rate at which a resource should be used. The optimum use rate is found where the benefits of using the resource minus the social opportunity cost is maximised. The second concept is that of total economic value. This looks at the component parts of the value of preserving a natural resource in a sustainable fashion. These component values will include the commercial or recreational use of the resource, its value to future users and its value in a sustained state to people who simply wish the resource to continue in existence. Both concepts are elaborated below.

The Social Costs of Resource Use

The particular feature of importance in social opportunity cost is to look at all the costs of using a natural resource. The costs will tend to comprise three components:

i)   the direct costs of extraction, harvesting or use;

ii) any costs that use of the resource now imposes on future users, the so-called 'user cost' element. For sustainably managed renewable resources such user costs are not usually significant since the resource regenerates (they are known then as 'stock effects' -- see Clark (1976)). However, if the resource is used non-sustainably a user cost does arise if the resource cannot be regenerated quickly in the future. For mixed renewable/exhaustible resources user costs can certainly arise, as with soils which are depleted severely. For habitats that are also non-renewable and for species which are depleted below their 'safe minimum standard', user cost also arises. For any future year, user cost is the difference between the costs that users now face and the costs they would otherwise face if the resource had not now been used. Total user cost is then the sum of these cost differences over time and discounted back to the present. Note that user cost, then, already embodies a discount rate.

iii) any external costs associated with use -- e.g. any adverse effects on soil quality, water, habitat etc., occasioned by the removal of tree cover, a specific water use, etc. and which effects are not 'internalised'. External costs will arise whether the resource is used sustainably or otherwise, but they will be particularly significant if the resource is used on a non-sustainable basis. Essentially, non-sustainable use means that the stock of the resource is being run down, and this is likely to give rise to more external costs.

The social opportunity cost concept can therefore be summarised as:

$$SOC = Harvest\ Cost + User\ Cost + External\ Cost$$

In symbolic terms we rewrite this as:

$$SOC = Ch + Cu + Cs$$

where 'h' denotes harvesting costs; 'u' denotes user cost and 's' denotes external costs.

The presence and magnitude of the cost components can be illustrated as follows:

	Resource Use:	
	Sustainable	Non-Sustainable
Ch	x	x
Cu	(stock effects)	xx
Cs	x	xxx

Total Economic Value

A natural resource has a consumption value. Trees are valued as raw materials, soil is valued as the agent of plant growth, water is valued for direct consumption or irrigation or other use. Wildlife may be valued for recreational purposes -- whether appreciation and/or physical consumption (e.g. wildfowling). Let us call this the 'consumption value' (Bc). Additionally, many people will wish to preserve the option to use a resource in the future. That option can only be preserved if the resource is preserved -- i.e. sustained. This value is the option value (Bo). There is also evidence for a third component of value -- existence value (Be). This value arises from people wishing to preserve a resource in a sustainable state because they value its existence without wishing to use it (consumption value) or reserve the option to use it (option value). These three component values comprise total economic value (TEV). [For further discussion see Pearce and Markandya (1987)]. Thus,

$$TEV = Bc + Bo + Be$$

The Relationship Between SOC and TEV

If a renewable resource is used sustainably its SOC of use is likely to be determined by the costs of harvesting and any external costs. (We are ignoring what are known as 'stock effects', simply for convenience). By definition, an exhaustible resource cannot be used sustainably -- its stock must decline over time at any positive use rate. For an exhaustible resource, then, SOC comprises the three cost components Ch, Cu and Cs, and its use is non-sustainable. Now, if a renewable resource is used non-sustainably -- i.e. the harvest rate exceeds the natural or managed yield of the resource -- then it is likely that costs Cu and Cs will also arise.

We can now identify the first source of resource management inefficiency. If there is no incentive to take account of user costs and external costs, or if they are inadequately accounted for, then there will be a tendency for over-use.

This conclusion has to be related to the relevant property rights regime. It holds, for example, if we are comparing outcomes under an open access regime (i.e. open access use rates will be excessive compared to an optimal open access use rate), or if we are comparing outcomes under a single owner regime. But single ownership regimes will tend to use the resource less intensively than an open access one, suggesting that 'privatisation' is one possible policy measure to conserve a resource. The form that a revised property rights regime might take is discussed in detail in the chapter by Paterson (Chapter 3). It suffices here to note that such regimes have to accommodate not just the kinds of readily identifiable uses that Paterson discusses, which are basically current water withdrawal uses by water authorities. While he argues toward the end of the paper that the principles involved are readily extendable to other uses, the discussion of total economic value above raises the issue of whether the redefined property rights solution is in fact as elegant and obvious as Paterson argues.

Thus, the externalities in question are not those arising from over-ascription of rights to an existing source, but from the activities which use the resource. How far they can be accounted for simply by defining rights to a source is very much open to question, unless we are to believe that the solution will emerge from bargains between users à la Coase (1960). This is more than unlikely as the literature since Coase has pointedly demonstrated. Moreover, it is unclear what source rights have to say about values such as existence value -- is some representative agency, say the government, to bid for these? Whatever the case, simply presenting the issue as one of redefining rights begs the question as to what a just distribution of rights is.

External costs are likely to be ignored for all the reasons that we are familiar with in the context of pollution problems. User cost will be downgraded the higher is the discount rate applied to the use of the resource. Private owner discount rates may well be above those of society in general, leading to excessive use rates (1).

The policy relevance of the concept of total economic value can also be seen in the context of a non-marginal decision to develop or preserve a wilderness area, say a wetland. Suppose the issue is whether or not to drain the land for agricultural use. Strictly, the decision should be based on a comparison of the costs and benefits of drainage. For drainage to be socially worthwhile we require:

$$[Bd - Cd - Bp] > 0$$

where Bd is the agricultural benefit arising from drainage, Cd is the cost of drainage and Bp is the foregone value of benefits of the wetland (we have ignored any costs of preservation). Notice immediately that this requirement for drainage to be worthwhile is stricter than the purely private decision in terms of the farmer's costs and benefits -- i.e. comparing Bd and Cd only. Bp is, in effect, the external cost of the development. Since it will be ignored under most market conditions we immediately establish that, under free market conditions, there will be a tendency for the overdevelopment of environmentally valuable land held in private ownership.

Now Bp in this formulation is in fact measured by total economic value (TEV). In short, we can rewrite the condition as:

$$[Bd - Cd - TEV] > 0$$

The important issue here is that TEV may be considerably larger than typical preservation values based only on rates of recreational use. By ignoring or playing down the option and existence components of TEV, then development decisions will be biased towards overdevelopment. The same will be true for issues of preserving species habitat -- too little habitat will be preserved. The analysis can also be used for issues such as hedgerow removal -- Bd would be the alleged gains to agricultural productivity from hedgerow removal, TEV would be the value of the wildlife and aesthetic losses.

TEV and SOC are linked. We saw that if a resource is used sustainably the external costs of resource use are likely to be less than if the resource is used non-sustainably. If it is used non-sustainably the stock of the resource will decline. In so far as this threatens (generally) non-commercial uses such as recreation, future use values (option value) and non-use values (existence values), it will give rise to losses in TEV. That is, TEV enters into the SOC formula as the obverse of external cost.

A further source of inefficiency in resource use can arise from government intervention in markets. Thus, subsidies may exist which accelerate the non-optimal depletion of exhaustible resources or the non-sustainable use of renewable resources. To the inefficiency that may arise because of the neglect of external environmental costs by the free market, may be added this government-inspired inefficiency, although the interaction of the two sources of inefficiency is not always additive (2).

V. TYPES OF RESOURCE MANAGEMENT INEFFICIENCY

The preceding section can be summarised in non-technical terms.

i) External costs and costs to future users arise whether a resource is used sustainably or non-sustainably, but they will be far more significant in the latter case.

ii) Different property rights regimes -- e.g. private ownership versus common property ownership -- can be compared in terms of their relative efficiencies of use.

iii) As soon as a renewable resource is used non-sustainably, its stock will tend to decline and this is likely to give rise to losses in total economic value (TEV). TEV is comprised of values reflecting use values, option values (willingness to pay to preserve the option of using the resource in the future) and existence values (willingness to pay to conserve a resource without making use of it).

iv) Because of the large potential in market systems for the neglect of preservation values (total economic value of the preserved resource), there is a clear bias towards the conversion of environmentally valued land for development purposes.

v) Government intervention in natural resource markets often has the effect of magnifying the resource inefficiencies identified in private markets.

vi) Redefining property rights clearly offers significant potential for minimising the costs of conflict of use in multiple use resources, as Paterson shows in Chapter 3. How far it accommodates the sources of value which properly make up a social objective function is very much open to question, however. Significantly, attempts to adopt this solution do appear to require a narrowing of the objective function to largely limited bureaucratic goals. This avoids the extensive 'fuzziness' that arises from standard neoclassical definitions of the objective function, but at the cost of possibly ignoring significant value components.

VI. MARKET FAILURE AND GOVERNMENT FAILURE

Despite the reservations of some Workshop contributors about the use of the distinction between market and government failure, it seems appropriate here to use it if only for organisational purposes. The exact way in which the simultaneous existence of market neglect of externality and government intervention affects resource use cannot be known with certainty. The reason for this is that resource depletion may well be less under conditions of imperfect markets -- i.e. 'monopolies'. That is, by and large, we would expect monopolistic resource owners to restrict the rate at which a resource is extracted or harvested. This is because they can make profit gains from such activity. Thus, we really have three sources of potential inefficiency in resource use which do not work in the same direction. These are:

i) externality -- the neglect of spillover costs from resource use. This will tend to make actual resource use rates <u>too high</u>.

ii) monopoly -- the restriction of output for profit reasons. This will tend to make actual resource use rates <u>too low</u>.

iii) government intervention -- the use of subsidies and tax laws which make actual resource use rates <u>too high</u> unless they are specifically designed to correct for (i).

In effect, we have an exact analogue of the situation for <u>pollution</u> where we know that a pollution charge that only takes account of the externality, for example, may produce non-optimal results if there is significant imperfection in the polluter's product market.

VII. EVALUATING SOURCES OF INEFFICIENCY

It is thus important to identify the sources of inefficiency and, as far as possible, to see which are the more important. This task requires a methodology for valuing the scale of inefficiency. As far as <u>externality</u> is concerned, it is worth pursuing the process of <u>monetary valuation</u> as far as possible. This might be carried out using the <u>recent OECD work</u> on benefits assessment as a basis, and in particular concentrating on aspects of total economic value, as described above. In particular it will be necessary to secure some 'feel' for the scale of recreational and wildlife gains that might be achieved by a more diverse species policy in forestry. Similarly, the costs of a forestry species diversification policy in terms of foregone commercial output must be estimated.

In terms of <u>government inefficiency</u>, recent OECD work and the Workshop paper by Frederick (Chapter 2) suggest that <u>water</u> charging policy frequently fails to recover development costs and also <u>fails</u> to reflect environmental costs of water supply. The Workshop paper by Sedjo (Chapter 5) shows that this is also true for <u>forestry</u>. This suggests a review of the various incentive mechanisms in place for water and forest management in OECD countries. Pursuing the government intervention aspect, it will be essential to investigate what financial incentives there are, in addition to market motives, which lead to externality-creating activities such as the drainage of wetlands. Government policies designed to reverse such activities also need to be evaluated. The same applies to habitat modification through development of land -- there exist incentives both to develop land and regulations to preserve such land, such as zoning. What is the net effect of the policies in place?

Other methodologies designed to elicit the scale of resource use inefficiency include <u>natural resource accounting</u>, <u>environmentally-adjusted national income accounts</u> and <u>resource assessment and monitoring</u>.

Natural Resource Accounting (NRA)

NRA is in existence in Norway and France (see OECD, Information and Natural Resources, 1986) and consists of various procedures for tracing flows of materials and energy within the economy, together with indicators of the physical state of resources. In effect, the accounts attempt to record the 'state of the environment' in the sense of looking at the quality of receiving environments, the demands made upon them, and the stock of reproducible and exhaustible resources.

The main attractions of an NRA system are:

i) it serves to monitor the effects of economic progress on the natural resource base;

ii) it provides basic information for use in planning at a broad level -- e.g. if stocks of selected wildlife or commercial resources (e.g. fish) are declining and appear to give cause for concern, there will be implications for overall land-use policy;

iii) it can assist in anticipating demands for resources and the problem areas that may arise.

At the most fundamental level, macroeconomic planning should have regular cognizance of the availability of natural resources and the impact of macroeconomic policy on those resources. NRAs provide a basic input in this respect. NRAs do not offer methodologies for evaluating the importance of resource stock changes.

Environmentally Adjusted National Income Accounts

Somewhat more ambitious in aim is the adjustment of existing national income accounts to reflect environmental costs. Various attempts have been made to do this and controversy surrounds them. Several purposes may be served by such accounts:

i) to measure more accurately the 'real' rate of development within an economy;

ii) to indicate the importance of environmental deterioration which would show up in increasing divergencies of 'adjusted' GNP from 'conventionally recorded' GNP;

iii) the obverse of (ii), to show how environmental improvement contributes to social welfare.

The disadvantages relate to the complexity of making adjustments in the same units as existing national accounts, i.e. money.

Resource and Assessment Monitoring (RAM)

RAM is similar to NRA but is likely to include additional information in the form of 'mapping' of resource trends making use of aerial and space photography. In many ways, it is better to think of RAM as an input into a system of NRAs.

The questions that arise in the context of such information systems are:

i)   what is the value of the information presented?

ii) to what extent do such systems assist in improving natural resource management?

VIII. POLICY INSTRUMENTS

The sequence of this overview so far has been:

a)   measure, as far as possible, the scale of resource damage;

b)   identify sources of natural resource management inefficiency;

c)   quantify, as far as possible, the importance of the inefficiencies in terms of:

   -- monetary values of costs of inefficiency, or value of gains to be obtained from improved management;

   -- non-monetary indicators through NRAs and/or RAM systems.

The rest of this chapter is concerned with the policy instruments that are available for improvement of natural resource management. A possible classification is as follows:

(1)   LAND USE REGULATION

The zoning of land that has multiple uses is a traditional and widespread procedure in OECD countries. Policies include 'green belt' zones round large towns and cities, development restrictions for sites of scientific or aesthetic interest, access restrictions to common land, creation of national parks, management of ecologically important areas by government agencies or quasi-government agencies. Given the very large extent of such policies it would seem sensible to focus only on certain areas. These might be classified according to some concept of 'ecological importance' -- e.g. wetlands, heathland, mountain reserves, and sites harbouring rare species.

At the simplest level of economic analysis it can be argued that land use regulation is inefficient because it does not permit the highest 'value in use' to occur. However, the values that determine this market solution are only those that can be reflected in the market process. If components of

21

economic value such as existence value are important they may not be expressible through market systems. In some cases they will be, however, and bids may be made to keep land from, say, a development use precisely because individuals wish to preserve it. A useful investigation would be one that quantified the extent to which land is purchased by individuals and groups concerned to preserve habitat and common recreational area. In short, what are the reasons for the public provision of areas such as national parks as opposed to private provision? One dominant argument is that certain land areas have the characteristics of public goods, i.e they can be jointly enjoyed by different people, and it is difficult to exclude people from them.

(2) RESOURCE PRICING

Natural resources are frequently priced at less than their marginal cost to society. Indeed, they may be priced below the marginal cost of production. This is because they are often distributed through public agencies which may have no incentive to break-even or maximise net financial returns. Water resources are a case in point. Elsewhere, subsidies may be given deliberately as with the exemption of agricultural land from taxes that apply to other land ('derating') in the United Kingdom. The value of this derating has been estimated at around £320 million in 1982/3 (3).

The effect of subsidies is to expand use of the resource beyond that which is optimal (unless it is judged that there are external benefits from so doing). Recommended pricing policy would involve some form of marginal cost pricing (short or long run marginal cost -- a dispute exists on which is most appropriate).

Note that resource prices can be varied according to use rates. Just as peak load prices are applied in, say, electricity, so prices can be varied according to peak demand for national parks, providing pricing is possible at all.

(3) OUTPUT PRICING

Natural resources, such as water, are often directly consumed. They are also inputs into the production of other goods, as with land for agriculture. If the outputs are subject to policies which encourage overproduction, too much land will enter into the agricultural sector. Many countries practice price support policies for the farming sector, resulting in overproduction of agricultural produce. The end result is mismanagement of land resources since too much land is devoted to agricultural use, in turn generating problems of loss of habitat and fertilizer runoff. United Kingdom farmers were estimated to have received overall subsidies of £2.8 to £3.7 billion in 1979 under the Common Agricultural Policy of the European Common Market. Removal of the subsidies would permit market forces to operate to the likely benefit of the environment. This proposition, does, however require testing.

(4)   RESOURCE TAXATION

An obvious instrument to use to control the rate of resource use is a tax. For exhaustible resources a tax which slows the rate of depletion is a severance tax which is a tax on the value of the output of the resource. This should have the effect of raising the value of the resource in situ, thus encouraging slower depletion. For renewable resources a tax is frequently advocated for common property resources so as to reduce use or harvest rates from levels where total cost and revenue are equated to levels where rents are maximised. Moreover, the government may then capture all the rent produced by the resource without altering the optimal level of activity. The tax may be on the harvest or use rate, or it may be on the level of inputs. Which tax is favoured in theoretical terms depends on the nature of the inputs, the stock of the resource and levels of uncertainty. Such taxes may have two components: one part which adjusts for any effect which current harvesting or extraction has on future yields (called the 'stock effect' in a fishery context), and the other which reflects the external costs imposed by any one user on another in a common property context (the 'crowding' effect).

The problems with tax solutions are familiar from the debate about pollution taxes. First, tax rates need to be estimated accurately at the outset since they are not always easy to change. Second, if they are wrong, output of the renewable resource can be reduced to levels which could cause major disruptions in the resource industry and even in the resource stock. Third, taxes on inputs (as opposed to outputs) face problems of definition and evasion: inputs that are not taxed can be substituted for inputs that are taxed. Fourth, there is a general political antipathy to 'new' taxes, a factor which partly explains the limited existence of pollution charges. Nonetheless, taxes are attractive instruments since they tend to avoid the bureaucracy and the costs of compliance that accompany regulation.

(5)   QUOTAS

Quotas are frequently used to regulate fishery catch rates. Are they equally applicable to other natural resources? In the fishery case the attraction of a quota is that it enables specific 'thresholds' to be observed. If there is thought to be some 'safe minimum stock' (SMS) of the resource, quotas could be implemented which prevent the stock going below the SMS (together with some margin to allow for uncertainty). Effectively, the quota permits future choices to be held open, which amounts to preserving the option values discussed previously.

Thus, the attraction of the quota as an economic instrument is that it is particularly suited to cases where the resource itself is threatened with irreversible damage. Quotas on use may of course be zero: the use of the resource is effectively banned. One problem with quotas is the manner in which they are administered so as to incorporate some kind of 'equity' between users. A total use quota with no rules of allocation will simply result in users rushing to exhaust the quota for themselves, with congestion externalities arising. The second problem is that, while a quota on a fish catch makes sense, it is not so obvious what a quota on use rates for ecologically sensitive habitat would mean. For example, are individuals to be rationed according to the number of times they visit the site? Familiar solutions are to close the site completely during important seasons -- e.g. migration times, certain weather conditions, and so on. Other solutions are available, as with the marketing of quota rights (effectively, marketable resource permits). Very simply, a market would emerge if quotas were made transferable and those attaching the highest use value to the resource would bid the most. Similarly, those concerned to reduce use further still could bid for the quotas and then not use the quota. The authorities can regulate use by buying-in or selling quotas through 'open market operations'.

(6)   PAYMENTS TO AVOID RESOURCE USE

It was noted above that those who wish to preserve resources could bid in the open market to buy the resource and then hold it without use. There is an obvious asymmetry in bargaining power between such purchasers and those who buy the resource for use. The former are unlikely to have any revenue flow associated with their purchase: by definition, keeping a resource out of use means spending money to buy the resource but receiving 'psychic' benefits only from the resource. If we think in terms of 'collateral', the purchaser who aims to use the resource has the present value of future net revenues to offset the cost of purchase. There are thus asymmetrical degrees of market power.

Partly for this reason, governments frequently act on behalf of the 'preservers' of resources. Apart from the regulatory approaches discussed previously, they may also intervene in the market by paying potential resource users NOT to use the resource. This provision is embodied, for example, in the United Kingdom Wildlife and Countryside Act of 1981. The general arrangement is that 'high value' sites (in an ecological sense) can be subject to management agreements between farmers and the Nature Conservancy Council (and local authorities for other agreements). Farmers can be compensated for profits foregone if they enter into a management agreement which involves not farming the relevant site. The perversity of the system is that much farm income is, in any event, subsidy, either through grants or through price support systems such as the Common Agricultural Policy. Effectively, compensation measures reflect inefficient input and output price regimes, thus raising the 'price' of conservation beyond what it would be if the system was efficient. If there is a given conservation 'budget' less conservation gets carried out than is justified.

(7)   TAX CONCESSIONS

Tax concessions apply to many areas of natural resource use. The fact that commercial forestry in some countries appears to be an undesirable investment when compared to rates of return achievable elsewhere has led to tax concessions being made which make the investment worthwhile. If the investment in question yields external benefits then such measures have economic justification. One problem, however, as far as the forestry example is concerned, is that commercial growing with the tax subsidy still only pays if the trees that are grown are conifers. The resulting 'monoculture' is problematic in many countries: limited wildlife species occupy conifer forests, they have limited recreational appeal, and the effect of closed canopies is to limit other types of biomass.

In the United Kingdom, for example, commercial woodlands are not taxed on actual income and expenditure but on a notional income which is one third of the assumed rent from unimproved pasture (in 1986 this made the effective tax base about £1 per hectare). Various other tax provisions are in place. Moreover, depending on the form of taxation which the owner may be subject to, grants in aid of planting can be obtained. Two notable effects emerge. First, investment in forestry in the United Kingdom is unrelated to the underlying profitability of the investment -- it is determined entirely by the grants/tax system. Moreover, it is very favourable as a form of investment the higher is the marginal rate of tax of the investor. Second, although grants are higher for planting broadleaved trees, establishment costs are also higher. The net effect of the tax and grant system is to make broadleaved investment relatively unattractive compared to monocultural conifers [Johnson and Nicholls (1986)].

(8)   REDEFINING PROPERTY RIGHTS

Prompted by the stimulating paper by Paterson (Chapter 3) a good deal of the Workshop was preoccupied with the property rights solution to resource problems. The main arguments have already been rehearsed, and the reader is urged to read Paterson's contribution. There he defines the essential ingredients of a property rights package designed to maximise some unstated objective function. The reader must satisfy his or herself as to the extent to which such redefinitions, if they are feasible (and Paterson argues strongly that they are frequently much more feasible than we might imagine), account for the rich variety of demands on a natural resource.

NOTES AND REFERENCES

1.   An expression for user cost is:

$$Cu \;=\; \frac{Pb.T}{(1 + r)^T} \;=\; Ca.T$$

where Pb is the price of the 'backstop' technology in year T, and r
is the discount rate. Hence a high value of the discount rate, r,
will lower the value of the user cost, Cu.

2.   A typical diagrammatic analysis is as follows. MEC is marginal
external cost and MNPB is the marginal net private benefit function
of the resource user. Qs is the optimum, but Qp is the amount of
resource use if the user ignores the externality. A subsidy can be
thought of as shifting the MNPB curve outwards to MNPB', making the
private optimum move to Qg, even further from the optimum. But the
diagram assumes perfect competition. The second diagram shows the
effect of non-competitive conditions. MR is marginal revenue, MC is
marginal cost, P is price. The true optimum is at Qt. Qf is a false
optimum. A subsidy applied in non-competitive conditions may thus
have the effect of moving Qf in the right direction.

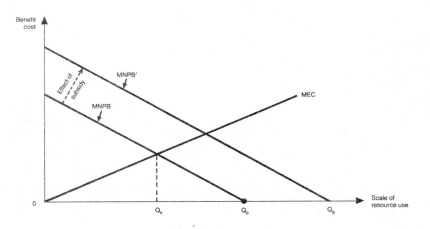

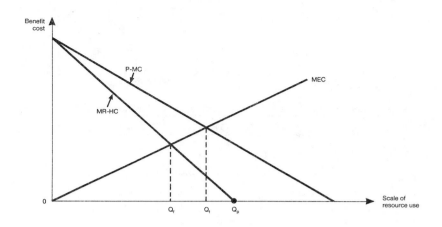

3. C. Clark (1976), <u>Mathematical Bioeconomics</u>, Wiley, New York.

4. R. Coase (1960), <u>The Problem of Social Cost</u>, <u>Journal of Law and Economics</u>.

5. D.W. Pearce and A. Markandya (1987), <u>The Benefits of Environmental Policy</u>, OECD, Paris, mimeo.

6. J. Johnson and D. Nicholls (1986), <u>The Value of Fiscal Measures to the Private Woodland Owner in Britain</u>, Department of Land Economy, University of Cambridge.

Chapter 2

WATER RESOURCE MANAGEMENT AND THE ENVIRONMENT:
THE ROLE OF ECONOMIC INCENTIVES

Dr. Kenneth Frederick

I. INTRODUCTION

Concern over the availability of good quality water is mounting in OECD countries. In the United States, water has been widely hailed as the likely source of the next major resource crisis.

That this should be true of water, one of our most plentiful resources, is somewhat surprising. Renewable supplies alone far exceed consumptive use, much larger quantities are stored in surface and underground reservoirs, and supplies seldom depend on maintaining the goodwill of foreign exporters. Moreover, water prices offer no hint of real scarcity or impending problems. Users commonly pay nothing for the water itself. They pay only for transporting or treating supplies, and even these costs are likely to be subsidised.

The anomaly of low prices in the face of rising concerns over supplies suggests a principal cause of some water supply problems as well as a possible solution. Users have little incentive to conserve or preserve resources which are virtually free. Indeed, the resulting incentives encourage misuse and abuse.

Higher water prices alone, however, will not resolve all our water problems. Some of the more serious threats to supplies are attributable to activities which make no intended demands on supplies. Landfills and agricultural and mining operations, for instance, can contaminate aquifers, streams, and lakes, leaving them unfit for many uses. The multitude of ways water resources are affected by human activities and the varied water quality requirements for uses as diverse as drinking, swimming, and fishing make it virtually impossible to know just what incentives would lead to something approaching an optimal use and development of water resources. Nevertheless, in view of the enormous distortions in the incentives now governing use of the resource, improved economic incentives must be part of any effort that hopes to resolve our long-term water problems.

II. ECONOMIC INCENTIVES

OECD countries rely on markets and market determined prices to allocate most scarce resources and goods. The market system generally works very well -- prices provide the necessary signals and incentives to encourage conservation and development of new supplies and to direct scarce resources to their highest-value uses. Water, however, is an important exception. Water markets are relatively rare and, where they do operate, the resulting prices often do not reflect the full costs of using the resource or provide reliable guidance for water conservation and development decisions.

Water poses special problems for establishing effective markets. Well-defined, transferable property rights are a necessary precondition for the smooth operation of markets. Moreover, if markets are to produce socially-efficient exchanges and appropriate price signals, buyers and sellers must bear the full costs and benefits of using or exchanging the property. Markets for water resources generally fail on both counts. The nature of the resource creates problems both for establishing the necessary property rights and for ensuring that the costs and benefits are fully internalised to the parties of a transaction.

Natural market failures, described in the next section, account for many but not all of the distortions in incentives. The policies and institutions which have emerged to fill the roles normally performed by markets also contribute to the misuse and abuse of water resources. Institutional failures are described in a subsequent section.

III. MARKET FAILURE AND THE NATURE OF THE RESOURCE

Several types of natural market failures make it difficult to create the conditions required for effective markets in water. Water may be a common (as opposed to a private) property resource, water use or transfers often affect individuals other than the user or the buyer and the seller, and water resources produce public as well as private goods.

The hydrological cycle makes water a renewable resource and accounts for its enormous capacity to assimilate much of a society's wastes. But this cycle also makes water fugitive in time and space. Rivers, streams, and groundwater supplies that flow from one property to another have the characteristics of common property resources for which it is difficult to establish clear property rights. In the absence of such rights, individual users face only part of the costs of their water use. And when ownership of a common property resource such as groundwater is established only by capture, an individual's incentive is to accelerate use of the resource before it is captured by others.

Water use is rarely fully consumptive. Consequently, water often gets reused many times. Each use, however, is likely to affect the quality, quantity, location, and timing of the water available to others. In the absence of special incentives or restrictions water users will not voluntarily take these impacts into account.

Establishing effective markets in water resources is also complicated by the fact that parties other than the buyer and the seller are likely to be affected by transfers among users. These externalities, or third-party effects, become particularly important when a transfer alters the point of diversion or return flow. Third-party effects must be taken into account to ensure that a water exchange does indeed result in an improved allocation of the resource. Neither the buyer nor the seller can be expected to represent the interests of other parties. Thus, when third-party effects are significant, some government oversight of water transfers may be needed.

A further complication in the use of markets is that water produces public as well as private goods. Public goods such as the scenic amenities associated with a pristine water body and the wildlife habitat they provide are not marketable because those who choose not to pay cannot be effectively excluded from enjoying the benefits. The incentive to free-ride results in less than the socially optimal level of investment in public goods by the private sector.

IV. INSTITUTIONAL FAILURES

The distortions resulting from natural market failures create a potentially important role for the government. However, the problems in achieving appropriate incentives and investment levels through government policies, laws, water management institutions, and administrative decisions are formidable. Non-marketed outputs such as wildlife habitat are inherently difficult to value, which makes it difficult to determine how much the public should invest in these areas. A more serious complication for policymakers arises when water supplies are affected by activities which do not even involve any intended water use. The impacts of many land use activities on water supplies are often unknown in advance and occur over many years, but the evidence mounts that the impacts are likely to be adverse and can be disastrous.

Governments have a major impact on the use and development of water resources in all of the OECD countries. Unfortunately, the governmental impacts are not always positive. As is indicated below for the United States, the government is also in a position to aggravate the problems of inappropriate incentives.

A principal function of government is to establish the conditions under which property rights are created and ensured. States have jurisdiction over water not explicitly encumbered by federal law in the United States. Federal law, however, can be very constraining as well as the source of considerable uncertainty over water rights. For instance, the Supreme Court has ruled that when the federal government withdrew land for Indian reservations or for any other purpose, it also implicitly withdrew sufficient water from the public domain to accomplish the purposes for which the land was withdrawn. So far these water rights remain largely unquantified. But in the water-scarce western states, they are potentially very large and a threat to the rights granted under state law. In addition, environmental legislation granting the federal government broad authority over activities affecting water quality and the commerce clause of the Constitution granting the federal government jurisdiction over navigable waters raise further questions as to the rights of the states and individuals to use water resources.

State water law was initially based on the common law doctrine of riparian rights granting the owner of land adjacent to a water body the right to "reasonable" use of the water as long as the use does not unduly inconvenience other riparian owners. This doctrine grants no priority of use and requires all owners to cut back in time of shortage. Riparian rights, which are still the basis for water law in the relatively water-rich eastern states, preclude the introduction of markets in water by tying the water right to the land.

In the western states where streams are fewer and flows less abundant and reliable, the riparian doctrine has been abandoned or modified in favour of the doctrine of prior appropriation. This doctrine which grants water on the basis of "first in time, first in right" imposes the full burden of shortages on the holders of the most junior rights.

Potentially, the appropriation doctrine was an important step in creating clear, transferable property rights in water. But the nature of the rights and their transferability have been obscured by a variety of legal and administrative provisions. The rights do not grant absolute ownership but rather a right to use water for beneficial purposes. Rights are lost when water is not used beneficially, and sale of a right has been interpreted as grounds for losing a right. The beneficial use provisions underlie the "use-it-or-lose-it" attitude that characterizes many water users in the west and is anathema to water conservation. Other constraints on the transferability of western water include restricting use to specific locations and granting certain types of use priority in time of shortage. Although these provisions complicate and restrict transfers, most western states do permit water transfers that do not impair third parties. But because third-party impacts are common and often difficult to resolve, water transfers are likely to involve lengthy administrative or judicial proceedings.

Providing for instream uses is a major deficiency in the appropriation-doctrine states. Instream uses such as wildlife habitat and fishing often predate extractive water uses, but most states made no provision for or discriminated against rights for instream use.

The federal government also has an important impact on the use and availability of water through the planning, financing, and management of water projects. Federal agencies have been eager to tame and divert rivers to meet virtually any perceived water need. Under conditions of water scarcity, however, the impacts of federal water projects on the nation's water problems have been mixed. Political considerations often replace economic criteria in project selection, environmental impacts of projects tend to be ignored or understated, inefficiency and distorted incentives are commonly a byproduct of the pricing of and restrictions placed on the use of federally-supplied water, and important uses--especially instream uses--are likely to be ignored or undervalued in the planning and management of the project.

The activities of the United States Bureau of Reclamation, an agency established early in this century to promote the development of irrigation in the arid and semi-arid western states, illustrate these problems. Bureau projects provide full or supplemental irrigation for about 11 million acres. Reclamation law originally intended for irrigators to repay full construction costs exclusive of interest charges as well as operation and maintenance costs. In contrast to the original expectations, more than 90 per cent of the capital costs of the Bureau's irrigation projects have been subsidised. On some projects the payments by irrigators are not even sufficient to cover operation and maintenance expenses (Frederick, 1982, pp. 66-71).

In addition to the misuse of public funds, Bureau projects may also result in inefficient water use. Restrictions placed on the use of federally-supplied water tend to eliminate both opportunities for moving water to higher-value uses and incentives to conserve. By locking water into specific uses and locations, the Bureau has created favoured areas within the water-scarce west where water is viewed and treated as a free resource. This profligacy exacerbates the problems and increases the costs of meeting water needs in other areas.

The Bureau of Reclamation is hardly the only agency to distort incentives through its water-pricing policies. By allocating highly subsidised water in accordance with the water requirements of the crops each farmer plants, some local water agencies have virtually eliminated incentives to switch to less water-using crops or to adopt water-conserving technologies. Price averaging is commonly used by water agencies to justify investments in new sources of supply for which costs exceed consumers' willingness-to-pay for additional water. By blending in high-cost with existing lower-cost supplies, uneconomic additions to capacity are rationalised. The use of declining block rates by some utilities discourages conservation by reducing the amount large users pay for additional water. More extreme distortions in incentives occur in cities such as New York and Denver which do not even meter most residential water use, and perhaps the most common distortion stems from the absence of any charge for using public water bodies for waste disposal.

Subsidies have also distorted investments in wastewater treatment facilities. Federal subsidies for sewage treatment facilities, first introduced in the mid-1950s, were set at 75 per cent of capital costs by the 1972 amendments to the Clean Water Act. Not surprisingly, such largess had important impacts on how municipalities dealt with their sewage. A Congressional Budget Office study (1985, pp. ix-xv) concluded that the subsidies resulted in larger, more sophisticated plants, less local involvement and pressure to contain costs, longer construction periods, and less innovative reuse of effluents. The subsidy was reduced to 55 per cent in 1985, a step that was expected to reduce the capital costs for secondary wastewater treatment plants by an average of 30 per cent. While the 1987 Clean Water Act amendments provided an additional $18 billion for construction of sewage treatment facilities, they also called for phasing out the federal role by 1994.

Countless examples might be used to illustrate how existing economic incentives encourage misuse and abuse of scarce water resources and how improved incentives might contribute to improved water use. California's Westlands Water District, which provides a particularly interesting and timely example, is considered below.

V. THE WESTLANDS WATER DISTRICT

The Westlands Water District, located on the west side of California's San Joaquin Valley, has more than half a million acres under irrigation in the midst of one of the world's richest agricultural areas. Initially, the district's irrigators relied on groundwater. By the 1950s, however, declining water levels threatened the area's continued prosperity, and a long-term, renewable water supply was sought. The federal Central Valley Project's (CVP) San Luis Unit has provided such a supply since the mid 1960s. In recent years this Bureau of Reclamation project has supplied the Westlands with more than one million acre-feet of water annually from the Sacramento/San Joaquin Delta about 150 miles to the north.

Several factors make the Westlands an interesting and important case study of the role of economic incentives in water use. The water delivered to the district under the CVP is highly subsidised, salt build-up threatens the long-term viability of irrigation in the district, drainage from the district has produced a highly-publicized "environmental disaster" which is focusing attention on existing problems as well as creating opportunities for reform, and there are high-value alternative uses for the water the Bureau delivers to the district. These factors are described below to illustrate how inappropriate incentives contribute to present water problems and how incentives might be altered to help resolve some of these problems.

Water Subsidies

The Westlands has acquired notoriety as one of the more egregious examples of the profligacy and distorted incentives found in federal irrigation projects. The generosity of federal reclamation law combined with the actions of the Bureau of Reclamation have enabled the Westlands Water District to receive CVP water at prices which are so low they no longer even cover operation and maintenance costs. In recent years, the Westlands Water District has paid about $10 to $12 per acre-foot, less than 10 per cent of the unsubsidised cost of delivering the water to the district. The resulting annual subsidy has been estimated by the Natural Resources Defence Council (NRDC) to average $217 per irrigated acre. For the average farm operation in the district (which has never met the size limitations reclamation law places on farms eligible to receive federally-subsidised water), the annual subsidy is nearly $500,000 (LeVeen and King, 1985, pp. 2-4). The distortions in water-use incentives are even larger because delivery costs represent only part of the social costs of Westlands water use. The full costs should also include the opportunity costs of water in the Delta and the environmental costs that result from the district's water use and drainage.

Reclamation law intended to provide a significant subsidy by exempting irrigation projects from paying interest charges and setting long repayment periods. According to the NRDC's study of the Central Valley Project, the Bureau has provided additional subsidies to Westlands farmers by extending the repayment period beyond the authorized fifty years, misusing the law's "ability to pay" provision, entering into long-term contracts with no adjustment for inflation, constructing a delivery and drainage system to unauthorized lands within the district, reclassifying some of the resulting drainage costs to circumvent Congressional funding limits, and subsidizing the electricity used to power the water pumps. The NRDC study concludes that "the illegal subsidies resulting from these practices exacerbate the many failures of the current system, which have created a vicious cycle featuring wasteful applications of water to marginal land to grow surplus crops." (LeVeen and King, 1985, pp. 2-3).

Drainage Problems

All irrigation water carries salts, and the salt concentration increases as the water evaporates, is transpired by plants, or passes over saline soils. Salts in the water or soil inhibit plant growth and in extreme cases leave land useless for agriculture. Every irrigator contributes to rising salt levels, but few farmers suffer the full costs of their contributions. If drainage is adequate and water is not too costly, farmers flush the salts from the root zone by applying water in addition to that required by the plants. Leaching seldom eliminates the salinity problem; rather it transfers the salts to lower points in the drainage basin. Thus, salt management involves externalities as farmers pass on some of the environmental costs.

Salt management is a particularly serious problem on the west side of the San Joaquin Valley where the soils are saline and the drainage poor. To maintain productivity, these soils must be periodically flushed and salt-laden water must be removed through subsurface drainage.

The area's drainage problems have long been recognized, and federal authorization of the San Luis Unit called for the Bureau of Reclamation either to work with the state to construct a master drain for the valley or to build a drain to take the project's effluent from the San Luis Unit to the Delta. Unable to reach agreement with the state for a master drain for the valley, in 1968 the Bureau initiated construction on a drain for the San Luis project. About a decade later an 82-mile stretch of the drain terminating in Kesterson Reservoir was completed. Although Kesterson was designed to serve principally as a regulating reservoir along the drainage route and secondarily as a wildlife refuge, it became the repository for the Westlands' effluent. A combination of environmental concerns over the impacts the drainage would have on water quality in the Delta and lack of federal funding delayed efforts to extend the drain to the Delta (San Joaquin Valley Interagency Drainage Program, 1979, pp. 4-4 to 4-5).

It was recognised that the drain to Kesterson did not provide a long-term solution for Westlands drainage. Yet, for several years it served as a repository for the runoff from about 42,000 acres with no apparent ill effects. In 1983, however, the United States Fish and Wildlife Service first noticed that Kesterson was becoming a deathtrap for the wildlife it was supposed to harbor. All but the hardiest of the fish species disappeared from the reservoir and large numbers of birds were hatched with grotesque deformities. The cause turned out to be selenium, one of many naturally-occurring trace elements contained in the drainage water delivered to Kesterson. The mineral first entered the vegetation and small animal life of the reservoir and eventually increased to lethal concentrations as it moved up the food chain. In small quantities selenium is considered beneficial to humans but little is known about the effects of higher levels on human health.

When Kesterson became the nation's newest environmental disaster, the first official response was to discourage birds from nesting in the reservoir and humans from consuming its wildlife. Longer-term solutions for cleaning up the reservoir are still being studied. While the debate over what to do continues, in March 1985 the Secretary of the Interior was forced, under threat of criminal prosecution, to act to avoid further violations of the Migratory Bird Treaty Act. After getting the district's attention with a threat to cut off the water, the Secretary agreed to continue supplying water in return for a pledge that the district terminate its use of the drain by June 1986.

The California State Water Resources Control Board closed the reservoir and ordered that it be cleaned up by January 1988. Without recognising the Board's authority over the reservoir, the Interior Secretary directed the Bureau of Reclamation to voluntarily comply with the Board's order. Who will pay for the cleanup and how much it will cost are yet to be determined. Cost estimates for cleaning up the reservoir after isolating it from further drainage start at a little over $1 million to keep the reservoir wet all year in the hope that this would reduce the selenium to insoluble and immobile forms that would settle in the bottom as sediments without threatening the groundwater. Cost estimates of removing all contaminated material to off-site landfills are nearly $145 million (San Joaquin Valley Drainage Program, July 1986, p. 3). The Bureau intends to treat the clean up costs as part of the project's capital costs to be repaid by water users. The Westlands, on the other hand, denies any responsibility for the costs (Western States Water Council, 1986). But if the Bureau's record of recovering project costs from irrigators is any indication, taxpayers are likely to pay most, if not all, of the bill.

With the closing of the San Luis drain in June 1986 the district had to find alternative ways of dealing with their drainage. This internalised the major environmental costs of the district's water use, providing Westlands an incentive to develop low-cost alternatives for reducing the quantity and toxicity of the drainage. Some of the alternatives being explored are deep-well injection of drainage waters, concentration and removal of selenium from the drainage, and incentives for farmers to reduce the quantity of drainage that must be dealt with. Preliminary evidence suggests deep-well injection below 5,000 feet would cost between $164 to $189 per acre-foot (San Joaquin Valley Drainage Program, 1987, p. 1). Several methods, including the use of microbes that eat selenium and ion exchange, are being studied to detoxify the district's drainage (San Joaquin Valley Drainage Program, July 1986, p. 2). The feasibility of these methods will depend in part on whether the resulting selenium concentrate can be sold (as claimed by proponents of the microbial solution) or must be isolated in a landfill. To encourage on-farm recycling, the district will pay irrigators $25 for each acre-foot of subsurface tailwater recycled. In its first year, this programme encouraged recycling on about one-third of the 42,000 acres that had drained into the Kesterson reservoir (San Joaquin Valley Drainage Program, 1987, p. 3).

The Westlands Water District apparently has not given up on the idea of a drain which would enable them to dump their contaminated waste water off on others. As part of an agreement made with the Westlands Water District in the summer of 1986, the Bureau of Reclamation is under a good faith obligation to attempt to secure authorization and funding for a drain. The district would pay the operation and maintenance costs and contribute $5 million a year (not to exceed $100 million or 35 per cent) toward the capital costs of the drain. The agreement appears to be another manifestation of the Bureau's generosity to Westlands and their disregard for the role incentives could play in developing low-cost solutions to water problems. However, the prospect that such a project will ever be approved currently seems very remote. Even before the recent concerns over balooning federal deficits and the awareness of a toxicity problem associated with the Westlands effluent, the project to extend the San Luis drain to the Delta was stymied by cost and environmental concerns. Pumping the drainage over the coastal range for disposal in the ocean has been discussed as an alternative. But just the talk of such a plan was sufficiently disconcerting to coastal residents that the state now has a law preventing the use of Monterey Bay for such purposes. The high costs of constructing an external drain and the reaction that inevitably arises from any location proposed as a waste-disposal site suggest the Westlands Water District may be left to handle their own drainage.

A more elaborate proposal for dealing with the drainage problem has been suggested by the Environmental Defense Fund (1985). The EDF plan calls for reducing the amount of polluted water by providing farmers incentives to invest in more efficient irrigation systems. The remaining drainage would then be desalted through reverse osmosis. The water salvaged through desalting and saved through increased on-farm efficiency would be sold to help finance the project. The salty brine residual from the desalinisation plant would be used to produce electricity in solar ponds, and if market conditions warranted, the selenium and other trace elements could be separated out for sale. The overall EDF plan appears doomed by desalinisation costs in excess of $1,100 an acre-foot (San Joaquin Valley Drainage Program, March 1986, p. 1) and problems in generating power from the salty residue. Nevertheless, two elements of the EDF proposal -- incentives for improved on-farm water use and marketing of conserved water -- are likely to be components of any successful effort to deal with the Westlands water problems without large additional inputs of government funding.

Water Marketing Opportunities

The San Luis Unit physically integrates the Westlands into the state's overall water network. The Delta, which is supplied largely by rivers to the north, is also the source of water delivered to locations as far south as Los Angeles and San Diego under the State Water Project (SWP). In view of the growing concerns over the costs and environmental impacts of large water projects and the uncertainties as to how southern California will meet its growing water demands, it is noteworthy that, without additional investment in infrastructure, water from the Delta now delivered for irrigation in the Westlands could be used to augment supplies in southern California's coastal cities or to meet the water quality needs of the Delta itself. The Metropolitan Water District, which services California's southern coastal cities, is actively seeking additional water and is a potential buyer of some of the water delivered to the Westlands.

As part of their planning for the State Water Project, California made contractual commitments for the eventual delivery of 4.23 million acre-feet annually. Current capacity of the SWP is capable of supplying only a little more than half of this amount. In the absence of new storage, by the year 2000 SWP commitments will exceed capacity in a dry year by as much as 1.4 to 1.6 million acre-feet. Southern California's coastal cities, which are losing access to some of their former supplies and encountering problems in developing new supplies even at high cost, have looked to the State Project to provide for most of their growing water needs. But there are now real doubts as to whether or not the state will meet its commitments.

A plan to add 1.9 million acre-feet to SWP capacity at a cost of about $3 billion was authorized in 1980. The plan called for construction of a 43-mile aqueduct, the Peripheral Canal, to take water from the Sacramento River around the Delta for shipment south through the California Aqueduct. This project would provide water to the Metropolitan Water District of southern California at an estimated cost of $216 per acre-foot (Wahl and Davis, 1986). Even though it is considered the least-cost way of adding to SWP capacity, two years after authorization the project was overturned in a statewide referendum. In a further rejection of additional large-scale interbasin water transfers, the state legislature subsequently rejected the governor's proposal to increase the capacity of the SWP with a through-Delta water transfer. Although the costs of supplying water to southern California by a through-Delta scheme are estimated to be about 50 per cent more than those for the Peripheral Canal, cost was only one of the factors underlying the rejection. Concerns about the impacts further diversions might have on the quality and availability of water in the Delta and the river basins of northern California underlie strong resistance from that part of the state to proposals for exporting more water south. While northern California's water may appear abundant when viewed from the arid southern parts of the state, water is now viewed as a scarce and valued resource in the north.

One possible source of additional water for Southern California is to purchase some of the water now going to farmers. A possible supplier of water might be the Westlands Water District, which faces rising costs associated with drainage problems as well as uncertain markets for their output. The potential gains from transferring a permanent right to an acre-foot of water per year between Westlands and the Metropolitan Water District servicing southern California have been estimated to be $1,260 (Wahl, 1986, p. 13). This figure understates the social benefits of such a transfer because it makes no allowance for environmental impacts. A water exchange from Westlands to MWD would reduce drainage problems in the Westlands and it would reduce the pressures for additional water withdrawals from the Delta and northern California's rivers.

The absence of transferable property rights poses a fundamental obstacle to establishing markets in federally-supplied water. Federal law and Bureau of Reclamation practices limit use of such water to designated project areas regardless of how inefficient or environmentally-insulting the use or how valuable the alternatives might be.

Publicity surrounding recent events involving the Westlands and Kesterson have stimulated interest in making the changes required to permit sales of Westlands water. Congressman George Miller even introduced legislation in 1986 to permit such exchanges. Miller's bill, however, never made it out of committee.

Ironically, some of the strongest resistance to marketing federally-supplied water stems from an aversion to the idea of permitting farmers to profit from the sale of subsidised water. Congressman Miller's bill was sensitive to this issue and proposed to split the profits. Proceeds from a sale would go first to pay the appropriate share of the operation, maintenance, and capital costs of the San Luis project. If the district used the funds for solving drainage problems, the remaining funds would be split 75 per cent for the district and 25 per cent for the federal government. Otherwise, the split would be reversed. The federal share would be used to repay costs of the Central Valley Project, to retire land, and to establish a groundwater management research area (Miller, 1986).

A crucial flaw in the Miller proposal may be a lack of adequate incentives for farmers to sell water. Limiting subsidies to Westlands farmers is a laudable objective, but pursuing this through restrictions on water use simply stifles incentives to conserve scarce resources or to transfer them to their best uses.

The NRDC study of the Westlands Water District offered several suggestions for limiting future subsidies while providing incentives to conserve and perhaps sell water in response to its higher cost. The NRDC recommends prohibiting all new expenditures on CVP facilities until provision is made for full repayment of all previous CVP expenditures and removing subsidies from water supplied through new facilities. To facilitate adaptation to the higher costs, they would permit farmers and/or irrigation districts to sell their water. Even under the NRDC plan, however, incentives to sell would be curbed by a recommendation that "proceeds in excess of fair compensation" resulting from water sales should be designated for economic development in the agricultural areas from which water is transferred (LeVeen and King, 1985, p. 7). The NRDC proposal addresses a commonly expressed concern in agricultural areas that water sales to urban users will be detrimental to rural development because the seller is likely to invest or spend the proceeds of the sale elsewhere. But if the "fair return" does not provide farmers sufficient incentive to sell water rights, there will be no sales or proceeds to distribute.

VI. TOWARD IMPROVED INCENTIVES

Lack of incentives to conserve and protect water resources and obstacles to reallocating supplies are not unique to California and the Westlands. Under-pricing of water, uncompensated environmental costs associated with water use, and wasted opportunities for transferring supplies to higher-value uses underlie most water-supply problems in the United States. The roots of these problems lie in the laws and institutions that treat water as a free resource and restrict where and how it can be used.

Historically, planners focused on supplying water for offstream use with little regard for the impacts withdrawals might have on instream values.

Water demand, which was considered to be insensitive to price, was projected to grow roughly in step with population and economic growth. Demand projections were labeled and treated as requirements, and planners proposed new water projects that would provide supplies well in excess of demand under all but the most severe drought conditions.

The traditional supply-side approach to water planning worked well when supplies were abundant and did not have to be rationed among competing uses, when the costs of developing new supplies were low, and when the environmental costs of water use were insignificant. Although such conditions are now rare, the supply-side approach with its emphasis on structural solutions to water problems still dominates the thinking of most water agencies.

High-quality water is now scarce and valued almost everywhere. But scarcity is a condition of virtually all resources and not sufficient cause for anticipating future water crises. Scarcity, however, does imply a need to limit demands on the resource as well as a need for institutions that can reallocate scarce supplies in response to changing conditions.

The nature of the challenge as well as the seeds for a potential crisis are evident in the California situation described above. The environmental as well as the economic costs of developing new supplies for southern California are high. Yet, rising costs are not the principal obstacle the area must overcome to secure the water needed to meet future demand. This relatively affluent area can afford higher-priced water with negligible effects on living standards. The real threat stems from a combination of the area's growing inability to secure new supplies at any price and their failure to curb their own water use through prices reflecting the costs of developing new supplies. A willingness to pay high prices is no longer a guarantee of securing needed supplies. But this is not due to lack of water. The existence of large quantities of water going to low-value and environmentally-degrading uses in the Westlands Water District that could be diverted further south at very low cost is evidence that ample water could be made available to southern California. But as long as the institutions and incentives prevent such a transfer, these supplies will do little, if anything, to quench southern California's thirst.

The United States is on a course which could well lead to a number of local or regional water crises in the not too distant future. Indeed, as is evidenced by the case of Kesterson reservoir, the past has not been without problems. A rash of new problems can be averted by protecting and making better use of existing supplies, and improving the economic incentives guiding water-use decisions may be essential to achieving this end. Irrigators will not conserve and reduce environmentally-degrading runoff unless water costs make it profitable or restrictions make it necessary. Domestic users will not treat water as a suitably scarce resource if they are not charged in accordance with the amount used and its costs. Industry will not develop and introduce alternative ways of disposing of their wastes as long as they can be freely discharged into water bodies. And municipalities will not invest in cost-efficient sewage treatment facilities as long as most of the costs are paid by the federal government.

The policy needs for water are simplified by the strong complementarity among the objectives of promoting conservation, providing for growing high-value water demands, protecting water quality, and reducing costs. Conservation is often the least-expensive means of increasing effective supplies, and the incentives needed to encourage conservation are likely to provide an efficient means of achieving many environmental objectives and of satisfying competing water demands. Moreover, water marketing may help introduce needed incentives without having to introduce unpopular price increases. Even where water rights are granted free, the opportunity to sell or lease the right provides some incentive to conserve and transfer water to higher-value uses.

Economic incentives will help but they may not be sufficient to meet a society's water-quality standards. Improved on-farm irrigation, which can be encouraged by charging farmers prices that more nearly reflect the full costs of their water use, is the most cost-effective way of reducing salt loadings, agricultural chemicals, and other pollutants attributable to irrigation drainage water supplies in the western United States (Frederick, 1987). But for some contaminants, especially toxics, the prospect of reduced loadings may not provide sufficient protection. Or the level of incentives needed to encourage the desired response may be politically unacceptable. Consequently, the command-and-control approach plays a major role in achieving water quality objectives in all OECD countries. Germany has attempted to set effluent charges to approximate damages and to encourage certain levels of treatment (Brown and Johnson, 1984). But as in France, where effluent charges are also assessed on industrial discharges, discharge standards specified in government-issued permits are the primary means of achieving desired water-quality levels. Moreover, a principal factor in setting effluent charges in both countries has been to raise enough revenue to cover the costs of programme monitoring and to subsidise treatment plants (Bower and co-authors, 1981, ch. 1).

In spite of their limited role to date, we should not conclude that incentives introduced in forms such as effluent charges and marketable permits do not offer effective tools for meeting water-quality standards at acceptable costs. Moreover, as the costs of the traditional command-and-control approach rise precipitously, some of these tools are likely to gain acceptance among policy-makers.

The availability of high-quality water at the turn of a tap has come to be taken for granted and is now an integral part of living standards in OECD countries. Consequently, threats to the continued availability of such supplies are cause for real concern. Change in the policies and institutions which gives a bite to these threats is inevitable. Indeed, change is already occurring as the costs and risks of current practices become more evident. Water crises certainly hasten the process of change but exact a high social cost that might be avoided by adjusting the incentives contributing to the misuse and abuse of scarce water resources.

REFERENCES

1. Bower, Blair T., Rémi Barré , Jochen Kuhner, and Clifford S. Russell. 1981. Incentives in Water Quality Management: France and The Ruhr Area, Research Paper R-24 (Washington, D.C., Resources for the Future, March).

2. Brown, Gardner M. Jr., and Ralph W. Johnson. 1984. "Pollution Control by Effluent Charges: It Works in the Federal Republic of Germany, Why Not in the U.S.?," in Natural Resources Journal, Vol. 24, October 1984, pp. 929-966.

3. Congressional Budget Office. 1985. Efficient Investments in Wastewater Treatment Plants (Washington, D.C., U.S. Government Printing Office, June).

4. Environmental Defense Fund. 1985. EDF Letter. (New York, November).

5. Frederick, Kenneth D., with James C. Hanson. 1982. Water for Western Agriculture (Washington, D.C., Resources for the Future).

6. Frederick, Kenneth D. 1987. "Discussion of the Mineral Water Quality Problem from Irrigated Agriculture" in Tim Phipps, Pierre R. Crosson, and Kent A. Price, editors, Annual Policy Review 1986: Agriculture and the Environment (Washington, D.C., Resources for the Future).

7. LeVeen, E. Phillip, and Laura B. King. 1985. Turning off the Tap on Federal Water Subsidies: Vol. I., The Central Valley Project: The $3.5 Billion Giveaway (San Francisco, Natural Resources Defense Council, August).

8. Miller, George. 1986. "A bill to amend the Act of June 3, 1960 authorising the Secretary of the Interior to construct the San Luis Unit, Central Valley Project, and for other purposes", submitted in the House of Representatives with an accompanying Statement of Congressman George Miller May 22, 1986 (Washington, D.C.).

9. San Joaquin Valley Drainage Program. March 1986. Status Report No. 3 (Sacramento).

10. San Joaquin Valley Drainage Program. July 1986. Status Report 4 (Sacramento).

11. San Joaquin Valley Drainage Program. January 1987. Status Report 6 (Sacramento).

12. San Joaquin Valley Interagency Drainage Program. 1979. Agricultural Drainage and Salt Management in the San Joaquin Valley: Final Report Including Recommended Plan and First-Stage Environmental Impact Report (Fresno, California, June).

13. Wahl, Richard W., and Robert K. Davis. 1986. "Satisfying Southern California's Thirst for Water: Efficient Alternatives," in Kenneth D. Frederick, editor, Scarce Water and Institutional Change (Washington, D.C., Resources for the Future).

14. Wahl, Richard W. 1986. "Cleaning Up Kesterson," in Resources, No. 83 (Washington, D.C., Resources for the Future, Spring).

15. Western States Water Council. 1986. Western States Water Issue #657, December 19, 1986 (Salt Lake City, Western States Water Council).

Chapter 3

RATIONALISED LAW AND WELL-DEFINED WATER RIGHTS
FOR IMPROVED WATER RESOURCE MANAGEMENT

Dr. John Paterson

SYNOPSIS

The superficial manifestations of allocative inefficiency in water resource management stem from two fundamental causes. The first of these lies in generic discontinuities in the legal doctrine applied to facets of the terrestrial phase of the hydrologic cycle and in poorly defined boundary conditions for water management in its catchment context. The second cause lies in the specification of rights at the point of delivery, rather than at the source. An appropriate basis for specification of rights is developed in some detail. The application of economic instruments requires a consistent metric for definition of rights such that additivity conditions are met, and that two alternatives exist for exhaustively partitioning water rights: release-sharing or capacity-sharing. Capacity-sharing is demonstrably superior. The operational features of a capacity share system are developed in terms of conditions for well-specified property rights. Translation from a "native" system of rights to a capacity share system is conceptually undemanding. Effluent or return flow obligations of right holders can be formally separated from the un-attenuated volumetric appropriation or right, and in practical terms, this is best achieved within a hierarchical system of water entitlements. Capacity shares lend themselves to creation of equal standing and technical consistency in the specification of environmental and in-stream obligations vis-a-vis private rights to divertable resources. Excess investment has compromised allocative efficiency in many systems in the past but this phenomenon is on the way out. In the "grandfathering in" of a new system of water rights sufficient gains should be available to compensate all stakeholders. New attention to the legal and technical basis of proprietory interest offers great promise for economic improvements in water resource management, but beyond these efficiency gains, well-defined rights and efficent markets are, in themselves, value-neutral. Outcomes will depend on broader societal norms.

I. INTRODUCTION

Very diverse constitutional, statutory and institutional arrangements exist among, and even within, OECD countries. Hence it is more profitable to dwell on the universal features of water as an economic good than to attempt endless examination of the bizarre consequences of market and government failures with which the critical literature abounds. It does appear that many governments wish to do better with water management than in the past. That is why they perennially tinker with institutional arrangements. There are two major requirements for making fundamental improvement:

. identification of a systematic approach that is clearly superior to existing arrangements; and

. solving the political problems of transition.

The contributions of economists to water policy have mainly concentrated on reform of pricing arrangements. Inefficient taxing and charging measures have been ubiquitous, so this effort was well directed. Pricing continues to be a vitally important avenue for reform in most systems. Every professional economist is comfortable with the micro-economics of utility pricing. Where pricing reforms have been introduced, significant gains have been realised. However, the emphasis on pricing has often diverted attention from more fundamental underlying "causes" of inefficient water resource management.

This bias in scholarly attention has been remedied to a small degree in recent years by those who have approached water resource management from the viewpoint of the theory of environmental policy, reinforced by the new analytical insights that arise from emphasis on "property" attributes of the resource and their consequences for its management. Rather than lament past failures, or rejoice at "paradox trouvé," it is more constructive to go to the basic artifacts by which most developed economies manage their water resources, and to examine the root causes of inefficiency.

Water is something of a technical curiosity among scarce renewable natural resources. Like other environmental resources with distinctive characteristic features, such as fisheries and forests, water poses its own peculiar difficulties for law, economics and institutions. Since the physical properties of water are most unlikely to change, social arrangements must make the accommodation. It will be argued that it is not too difficult to design appropriate water management arrangements from first principles, and we are in the process of doing so in Victoria. The design problem becomes distinctly non-trivial when implementation of the solution must also satisfy a complex and internally contradictory set of existing entitlements, many of which while poorly defined are, or can be, stoutly defended in the courts or the legislature. It will be further argued that if improvements have real substance, their economic yield must be sufficient to compensate the losers who could otherwise exercise a veto on the introduction of these improvements.

II. CONSISTENT VESTING

The "property rights" school has brought valuable new perspectives to bear on water policy, but contributors have usually dealt only with fragments of the resource, such as divertible water with private economic value, rights to pollute or environmental and common property values. However, the origins of economic inefficiency in water resource management usually lie so deeply buried that mere improvements in the transactional framework will generally be insufficient to rectify them.

Most legal traditions associated with water have their origins in the settling of private disputes. Even when these are codified in statute law, as they often have been, the traditional terminology usually survives. Hence it is common for <u>facets</u> of the terrestrial elements of the hydrologic cycle to be dealt with <u>generically</u> under different laws. Different legal doctrine often coexists uneasily within a common legal framework, dealing inconsistently with:

. divertible surface water resources

. resource attributes of unconfined surface flows

. drainage or nuisance attributes of surface flows

. divertible groundwater resources

. environmental values in rivers, streams, lakes and wetlands

. water quality

. powers and duties of authorities.

Within and between each of these facets of water the boundaries between public and private rights, the legal standing of the public interest, and even the basic definition of the subject of the law may be haphazardly drawn. Spectacular instances of inefficient water resource allocation processes often divert the attention of the analyst from these more fundamental causes. Any system of economic incentives that rests on a disjointed foundation of basic law is inevitably second best and is highly productive of unforseen and unwanted consequences. We must look beneath surface inefficiencies to the basic legal foundations of water management if we are to establish a firm basis for the application of economic instruments. Inconsistencies in laws relating to the seven facets of water management listed above would be represented in any optimising tableau as technical constraints, and of course, each binding constraint has a cost.

The legal foundation of water resource management must capture the subject, that is the terrestrial path of the hydrological cycle, in a single vesting framework if these gratuitous constraints on management performance are to be relaxed or eliminated. The way this integrated vesting is accomplished will quite properly differ between jurisdictions, since legal and institutional reform in each society must respect its own precursors if reform measures are to win legislative support. That the foundations must be properly laid is beyond debate. No definition of proprietary interest is possible without a clear specification of the subject of that interest, and no specification of the subject is possible if a generic shift can alter its legal identity.

III. CATCHMENT MANAGEMENT: SEPARABILITY

In writing statute law to deal holistically with water, the elimination of internal generic fragmentation must be matched by well-conceived limits to the ambit of water law. The watershed or catchment setting ultimately embraces every physical artifact of human society. Unless strong separability can be established to define the outer limits of water management responsibilities, the theoretical potential arises for water law to pretend to a totalitarian tyranny à la Witvogel. More immediately, the "water" bureaucrats will be inextricably enmeshed in pervasive territorial conflict with their peers in the fields of land protection, agriculture and forestry, fisheries and wildlife, and urban planning.

A recent popular best-seller in the management literature [Peters and Waterman (1982)] made a previously well established point in vivid terms by pronouncing that excellent organisations "stick to the knitting". It is easy to manage well when a narrow and well-defined set of objectives is continuously before all members of an organisation. That circumstance was available to all of the legendary public sector organisations that built major highway and water systems or provided pioneering forest and soil conservation services. These public bodies commonly exhibited most of the characteristic features of excellent organisations because the conditions existed in which they could stick to the knitting. They knew what they were there for. They enjoyed broad political support and they had access to financial resources commensurate with their tasks. Their people developed high levels of technical expertise and displayed predictably high motivation and morale. They were not obliged to reflect on the external consequences of their onward rush. Cities fell before the freeway builders and natural environments were radically modified by the dam builders but that was not their concern. They knew what they had to do and they did it.

With the rise of the environmental movement, the previously simple settings of the bureaucratic dinosaurs suddenly changed. Popular support for their developmental vision became fragmented and they were asked to recognise a wider range of societal objectives. Co-ordination between government programs became a new cult, and perhaps properly so. However, the conditions which had existed for excellent organisation around a narrow purpose disappeared. People who previously could devote single-minded attention to building the next dam found their time dissipated in the unheroic setting of the co-ordinating committee, or in hearings controlled by a regulatory agency that had never built anything. The cost of getting things done increased. At the same time the taxpayers revolted at the increasing costs of government, while the compounding complexity of government operations bewildered and then enraged important constituencies. At the heart of all this was an attempt to respond to the environmentalists' truism that ultimately everything is connected to everything else, coupled with the "Comprehensive Rational Planning" movement in public administration.

The ideal of broadly consistent societal objectives and co-ordinated official action directed to their achievement has been quite widely embraced by well intentioned public officials since the 'sixties but the reach has consistently exceeded the grasp. There are fundamental reasons for believing that the rational comprehensive approach cannot ever be realised in practice. Consistent stable preference ordering in public choice appears to be an impossibility. Sufficient attention to major interactions between primary objectives of public bodies is to be preferred over attempts to deal comprehensively with all interactions (Simon 1983, Paterson 1986).

If public bodies are to perform effectively, then they must know what their mission is. That means narrow and well-defined functions. Life is made simpler for all concerned if formal conditions for functional separability are recognised in defining jurisdictional boundaries. In terms of the present topic some classes of water/catchment interaction are salient; they may be dealt with explicitly and assigned by statute to one or other source of administrative power: this class to farm administration, that to water, and so on. Where multiple heads of political authority have an interest as, for example, in control of point sources of pollution, a separate "arms length" agent of statutory authority, such as the EPA may be called for, to distinguish poacher from gamekeeper.

Diffuse and pervasive classes of localised interaction can be dealt with only in plans of management, since general purpose statutory powers can never fully specify the utility surface that might in principle determine the desired solution. Adequate treatment of "external" boundary conditions for water law is quite as important as the elimination of distortions created by artificial internal boundaries (Paterson 1986).

IV. A CONSISTENT SYSTEM OF MEASURING WATER

Contemporary water resource economists commonly emphasise the importance of rationalising "property" attributes of claims on water resources in the economic enhancement of management arrangements. The inadequacy of the non-priority "riparian" doctrine as a basis for enhanced definition of entitlements is generally recognised. The "prior appropriation" doctrine offers somewhat more promise because of the presence within it of a ranking variable, seniority. Seniority contributes a dimension in the title to a share of the resource that can match the sum of subordinate water entitlements to the stochastic profile of the bulk resource from which the shares are derived.

The contemporary literature on property attributes of water entitlements is long on normative content and a little short on basic technology. The importance of rationalising the property base is substantial but the benefits must not be oversold, as some (Anderson 1983) have tended to do. This point is pursued in Section XIV below. By shifting the focus from the normative to the positive some easy lines of advance come into view. All legal systems deal with surface water resources by distinguishing:

1) rights in non-confined or diffuse overland flows;

2) rights in natural concentrations of water; and

3) rights to supplies from public systems.

These are provided for in various ways, each, as a rule, in a way which reflects traditional doctrine in the particular polity. The first often rests on drainage doctrine and on attributes of land ownership. The second usually rests on a mixture of common law and statutory "water rights" doctrine. The third normally rests on statute law, corporate articles of association, consumer protection and regulatory agency law. They are rarely specified in terms that "add up" to 100 per cent of the resource in question. A metric is required which has the essential adding-up property. Without such a metric the superficial distortions so often lamented, and so often denounced from a normative viewpoint, will remain.

Among existing water allocations systems, and with very few exceptions, the main classes of right, whether to unconfined flows, to natural water sources, or to water from public systems, are <u>attenuated</u> to a greater or lesser degree. The concept of a non-attenuated property right (Cheung 1970, Randall 1975, 1978, 1981) requires that the right should be:

. explicit
. exclusive
. enforceable
. transferable

Market efficiency, and the applicability of economic instruments in water resource management, requires that rights in water should be unattenuated or in other words, that the underlying legal doctrine should embody these four characteristics. Much of the prevailing confusion about the economic character of water, that is, its spread across the public-goods/private goods spectrum, can be eliminated by constructive use of the four necessary conditions for non-attenuation.

The four necessary conditions can be met only if the instrument of entitlement permits, in principle, an exhaustive partitioning of the resource among title holders. <u>This can be achieved only at the source</u>. The primary reason that the essential simplicity of the answer has been overlooked lies in doctrinal traditions that define entitlements not at the source but at spatially distributed delivery points. Catchments exhaustively partition the topography and its water yield. Partitioning of run-off, of storage capacity shares and of (evaporative and seepage) losses simultaneously satisfies the requirements for:

. exhaustive and unique partitioning of the resource;

. a legal basis for a consistent system of non-attenuated entitlements.

This point is so basic to the application of economic instruments, and so generally neglected that detailed elaboration is called for.

V. WATER RESOURCE MEASURING

The aggregate of individual "paper" claims on a catchment must be equal to the aggregate divertable physical resource in the catchment if the paper claims are to serve as an adequate basis for efficient transactions in water. Economists and accountants are ever vigilant against multiple counting and non-additivity in national income and commercial accounting systems. Yet when the same disciplines examine water resource systems, basic arithmetic goes out the window. Intermediate and non-additive claims are either taken at face value, and a resource is simply said to be under or over-appropriated, or else they are taken to be so mired in downstream and return-flow obligations as to be completely indecipherable. Yet, in reality, the aggregate physical resource is always precisely translated into ex post shares which are exhaustive. Consequently non-additivity in paper claims is properly viewed as a government, or legal, or analytical failure. How well the market may be doing in each case of ex-ante claims that are at variance with ex post allocations is anybody's guess. A little attention to basic accounting is worth a tonne of high-powered analysis if we are looking for economic advances in water resource management.

All downstream entitlements to surface flows ultimately depend on shares of the resource at the point of origin, which may be a headwater storage, or unregulated runoff, or, more often, a combination of these. For simplicity, and for some other reasons which will be described in Section X below, let us consider only the "big" claims by a small number of supply authorities operating in a catchment. Suppose two water supply authorities share a common source of water. Each believes that it is entitled to a one-half share of the water regulated at the source. The question is, does this adequately define the rights of the two parties?

Suppose Authority A regularly takes one-half of the safe annual yield of the shared source while Authority B regularly takes less than its full share. Annual carryover in the storage will, as a result, be greater than that required for given target securities of operation by the respective authorities. The source would be described as fully appropriated but undercommitted. Now suppose a drought occurs. Winter inflow is well below average, and less water than usual is available for summer release. Authority A will be inclined to argue that it is still entitled to half of the (reduced) yield available. Authority B will argue that since it withdrew less than its full share in the previous year, it in effect carried over more than Authority A and hence is entitled to more than half of the reduced aggregate pool of water available for release. The specification of entitlements in terms of vague "50/50" shares gives no grounds for resolving the impasse. In effect Authority A is arguing that the 50/50 split relates to releases. Authority B is implicity arguing that it applies to the capacity of the storage, and inflow to the storage, or in other words to a capacity share. For the entitlement to be fully specified, the specification requires more parameters to give meaning to the vague "50/50" share.

If specification refers to releases then the rights of Authority B are attenuated because the security of supply of Authority B is vulnerable to withdrawals by Authority A: whatever water is carried over, even if it remains in storage purely because of rationing by Authority B, will be shared by Authority A. The benefits of "responsible" rationing can be "appropriated" by the Board that chooses not to ration.

If on the other hand the specification of a 50/50 split refers to stored water, then neither Authority has complete control over its security/yield choices. In each year, each Authority is entitled to one half of the residual stored water, which is partly determined by the previous year and current year use by each Authority. In the extreme case, if the shares are defined from month to month, the 50 per cent entitlement of Authority B "evaporates" almost completely, as Authority A draws down the pool and then claims half of the residual.

Hence it appears that there are severe limitations to any definition of entitlements in terms of releases of stored water. In theory it is possible to devise a mathematical formulation of shares in releases which is somewhat better than the naive rule proposed in our hypothetical example (Buras 1985). A third option is developed in Sections VI and VII below, which will be described as "capacity sharing", following Dudley and Musgrave (1986 and 1987, forthcoming).

Pursuing the example of our hypothetical shared water source, we saw the uncertainties created by entitlements that were not explicit. Both the release sharing and the stored water sharing rules failed to confer an exclusive advantage to the Authority that sought to protect the security of its own supplies. No secure entitlement to water could be enforced. Because water was continuously "up for grabs" there was no basis available for trading by negotiated transfers between supply authorities.

An instrument of entitlement is called for which is explicit, exclusive, enforceable and transferable. Authorities depending upon a shared system must be enabled to operate as if they each had exclusive tenure of equivalent catchment and storage capacities, by means of an appropriate instrument of entitlement. Such an instrument can in principle be defined in terms either of release shares or of capacity shares. The next step is to decide which alternative to choose.

VI. RELEASE SHARING OR CAPACITY SHARING?

The primary purpose of a reservoir is to regulate streamflow. Rainfall that arrives when water is not in demand is stored and "transported through time" for release when required. The stored water can be used for recreation, hydroelectricity generation, irrigation, stock and domestic use and town supply. Empty storage capacity or "air space" can be used for flood protection. Storage of water in a reservoir results in seepage and evaporative losses which reduce the yield of a regulated source relative to natural streamflow. In return for this loss of water, that water which remains in storage is able to be delivered, at times, and in amounts, that users value more than the natural arrivals. Investment in regulating works is warranted when the capital and operating costs of those works, plus the environmental damage cost entailed, are less than the additional value bestowed on the water by regulation. When development of a shared source is contemplated, and when the beneficiaries are responsible for meeting the full cost, rational investment choice demands that each beneficiary should be able to measure precisely and evaluate their respective potential benefits.

Release-sharing defines entitlements in terms of delivered water with a given volume and reliability. The owner/operator of the storage decides what amount is available for release in the light of circumstances, and what "prudential" carry-over is required to meet the obligation of the operator. Capacity-sharing allocates explicit shares of storage capacity, inflow, losses, and hence of stored water, to end users, each of whom decides, in regard to their own respective shares, what amount is available for release, and what to carry over.

On the face of it the distinction may appear a subtle one. We shall see, however, that the two possibilities lead to quite different approaches to specification of entitlements, and potentially very different practical results for authorities and for the end users who are their customers.

An accounting identity or balance equation can be stated, relating releases to states of the hydrologic system:

$$R_t = S_t + Q_t - L_t - S_{t+1} \tag{1}$$

where: R_t is the release from the reservoir in period t

S_t is the water in storage at the beginning of period t

Q_t in the water inflow during period t

L_t is the system loss (evaporation plus seepage) in period t

S_{t+1} is the water remaining in storage at the end of period t

Water inflow and system losses over the period of interest are random and depend strongly on seasonal factors. Water in storage at the beginning of the season of interest, S_t is determined by the operating policy in the previous period. The operating policy in any period is a rule for choosing either the maximum release available for that period, or the end of period carry-over. Specification of one of these two variables immediately fixes the other once assumptions are made about probable inflows and system losses for the period. Of course the actual climatic outcome for the period determines what actually happens. Release and carry-over decisions made prospectively at the beginning of the period are continuously revised as events unfold.

Operating policies can be of an ad hoc nature or can be determined rigorously. Rigorous operating rules are usually derived to optimise some measure of system performance. The measures of system performance most commonly adopted at present are physical in nature; (see for example Louckes et al (1981) or Buras (1985)) but in principle can reflect economic objectives. In fact a kind of "in between" practice often applies, which more or less reflects the present state of the art and is based on a probability analysis of low flows or "droughts of record".

The difference between capacity sharing and release sharing can be demonstrated using the water balance equation. Let Pi (i = 1,..., n) be the share of releases allocated to user i under release sharing, or the share of inflows, capacity and system losses allocated to user i under capacity sharing.

Under release sharing the water balance equation for user i can be written:

$$P_i R_t = P_i S_t + P_i Q_t - P_i L_t - P_i S_{t+1} \tag{2}$$

where the release in season t is determined by the reservoir operator.

Under capacity sharing , the water balance equation is:

$$P_{i1t} R_t = P_{i2t} S_t + P_i Q_t - P_i L_t - P_{i3t} S_{t+1} \tag{3}$$

where:

P_{i1t} is the proportion of total releases owned by capacity holder i

P_{i2t} is the proportion of start of season carry-over owned by capacity holder i

P_{i3t} is the proportion of end of season carry-over owned by capacity holder i

The proportions in (2) and (3) must satisfy:

$$\sum_{i=1}^{n} P_i = \sum_{i=1}^{n} P_{i1t} = \sum_{i=1}^{n} P_{i2t} = \sum_{i=1}^{n} P_{i3t} = 1,$$

or, in words, the total of all shares of each term in the balance equation must sum to unity. If total storage capacity is K, we must also have:

$$P_{i2t} S_t < P_i K \tag{4}$$

$$P_{i3t} S_{t+1} < P_i K \tag{5}$$

or, in words, the volume of stored water held by each shareholder at the beginning and end of the accounting period must be equal to or less than the physical volume of storage owned by each shareholder.

Equations (4) and (5) reflect the fact that the share of water in storage owned by capacity shareholder i at the start and end of each season cannot exceed that shareholder's portion of total capacity.

These equations do no more than formalise the accounting basis of shared operations. However, the flexibility of capacity sharing as compared to release sharing is evident from the arithmetic fact that capacity shareholders can actually control two variables (3) by determining $Pi2t$ and $Pilt$. Under release sharing, owners of releases control none of the variables in (1). This is the physical implication of the attenuated nature of the property right owned by participants in release sharing.

The algebra does no more than demonstrate the potential for independent decision-making by supply authorities that draw on a shared source. Independent control is possible, so we have the comfort of knowing that it is feasible in principle to create bulk entitlements that are non-attenuated, and which give each capacity sharer control over its own destiny. The analysis also demonstrates that release sharing is inferior to capacity sharing in this regard, because rights to release shares must inevitably be attenuated. Next we develop a framework for making a capacity share scheme operational.

VII. AN OPERATIONAL AND LEGAL BASIS FOR CAPACITY SHARING

A capacity sharing system can in principle provide a basis for non-attenuated bulk entitlements. Such a system requires explicit legal specification of the entitlement of each "bulk owner" in terms of:

(a) volume of storage, i.e. share of reservoir; the physical storage volume of the reservoir;

(b) share of inflow;

(c) procedures for calculating and allocating shares of losses; and

(d) precise measurement of "my" releases.

All of these are physical quantities that can be directly measured, or estimated with a high degree of precision. Each appears in the balance equation. Each is fully and consistently defined subject to the constraints (4) and (5) that no shareholder can store more water than that shareholder's portion of the physical storage.

Constraints (4) and (5) call for a rule, forming part of the legal instrument of entitlement that specifies the procedure to be followed when inflow occurs and one shareholder has no remaining storage share. There is no problem if all shareholders' capacity entitlements are full; water physically spills from the storage. But if one share of capacity is occupied, inflow simply occupies the vacant capacity of other shareholders. We will refer to this situation as an "internal spill" and add to the requirements of (a) to (d) above; a further requirement:

(e) rules for allocation of internal spills.

If the bulk entitlement is defined in terms of (a) to (e) we have satisfied three of the requirements for a non-attenuated property right. It is explicit: all terms in the balance equation are specified. It is exclusive, since, subject to the internal spills rule, the operating decisions of each shareholder are completely independent of the choices of others. It is potentially enforceable; all shareholders can precisely identify their own entitlements, so all that is required is legal vesting to establish enforceability.

It is necessary only to add rules for transferability in order to have achieved our objective; bulk entitlements that are non-attenuated. Hence, to the requirements (a) to (e) we add a further right to be associated with a capacity share entitlement:

(f) the right to transfer inflow, stored water or storage capacity between the parties, upon such terms as those parties shall agree.

A system of this kind will allow each right holder in a shared storage to operate as if each "owned" a dedicated free standing reservoir and catchment. All decisions are in the hands of the respective owners.

Within such a system, the reservoir operator acts as a kind of banker, maintaining a continuous account of the "balance" held by each customer. The operator deducts the amount of releases from storage shares, adds shares of inflow, deducts estimated losses, apportions internal spills, and turns on the tap when instructed by a "depositor" to make a release or "payment". The operator debits and credits the accounts of the capacity sharers when a transaction (i.e. a water transfer) occurs between them.

While such a system is markedly different in its nature from most existing arrangements, its introduction would, in most circumstances, be a straightforward matter. A "continuous accounting" system of this kind is at present under active consideration by the River Murray Commission for improving the management of water resources subject to the River Murray Waters Agreement in Australia, and variants exist elsewhere.

VIII. ILLUSTRATION OF CAPACITY SHARING IN OPERATION

Partitioning Storage Capacity

For any given reservoir, the active storage volume is a function of the water level between "dead" storage and spillway height. This volume/depth relationship is routinely calculated by the operating authorities responsible for most major storages in current use. In the simplest case, an active capacity of 100 units can be simply split into notional parts that add up to 100, to translate existing release rights into equivalent storage capacity units.

In practice the translation or partitioning into the terms of the balance equation will be a little more complex than this because:

- a user with a prior entitlement to, say 50 per cent of releases but with greater prior security of supply (say a town supply authority sharing with a "non-carryover" irrigation user) is implicitly entitled to more than 50 per cent of capacity;

- storage that is "dead" to one user, relying on a high level offtake, is "live" to another that is served by down-river releases from a lower offtake;

- temperature can be a pertinent consideration in environmental releases, and selection of offtake level is also used as a means of water quality management by town supply authorities. Hence prior capacity entitlement can be a function of the range of offtake levels available to the prior holder of a release share.

Partitioning Inflows

For the above reasons, the identification of capacity share equivalents of prior release entitlements cannot be left to some simple formula. In each case an agreed split must be arrived at to establish capacity shares. A legal basis for this must allow negotiation between the interested parties, followed by statutory recognition to create rights in storage capacity.

The share of inflow equivalent to a prior release entitlement will not necessarily equal the equivalent share of storage capacity. An existing release right that operates on the basis of large carryovers is implicitly receiving a smaller share of inflow in relation to shared capacity than one with no carryover, assuming that there is no significant difference in respective securities of supply. Again a formal procedure for establishing initial inflow shares, and for entrenching them in a legal instrument of entitlement is called for.

Partitioning Losses

Depending on the shape of the storage, losses vary according to aggregate water level and losses will generally be a positive (i.e. increasing) function of water level. To calculate the apportionment of losses between capacity holders who at any given time hold different notional levels in store requires explicit rules that have no counterparts in prior arrangements.

Entitlements to Internal Spills

When an inflow share exceeds the available "empty" capacity of a right holder, but the reservoir does not spill over, a rule for allocation of internal spills is required. The conceptually simplest rule would be to allocate an internal spill proportionately among all other right holders with spare capacity. Other possibilities exist and different methods for allocating internal spills might be adopted from case to case according to circumstances. Operational flexibility can be enhanced if the "operator" holds a capacity entitlement from which miscalculations can be rectified.

This is equivalent to the "fidelity fund" sometimes operated by those who hold money in trust.

Operating Responsibility and Costs

Under the kind of capacity sharing arrangement described, the role of the operator would change markedly from that which commonly applies at present. The operator becomes a manager who carries out instruction from share holders but who has no role in risk/reliability decisions. Reliability/yield decisions become the responsibility of individual shareholders. Their operational policies become substantially independent of each other, and of the reservoir manager. The role of the manager must be redefined in law and practice. Arrangements are required whereby the right holders each meet part of operating, maintenance and water accounting costs incurred by the manager. In the case of new projects, the share parameters would be defined in advance. In many cases the operating authority will be one of the shareholders, but the roles of shareholder and operator must be rigidly separated in law.

IX. EFFLUENT (RETURN FLOW) OBLIGATIONS

In any "native" sytem of water rights volumetric entitlements are established at spatially distributed points of receipt. Attenuation is inherent in this formulation, because it imposes effluent or return flow obligations on upstream diverters. The riparian doctrine confines entitlements to "ordinary uses" which may be determined by the courts. Diversion under a riparian entitlement is limited to such an amount that the flow in the stream "is not sensibly diminished". This judicial formulation is the despair of economists who are inclined to view riparian rights as hedged in meaningless legal jargon.

Economists have been a little more comfortable with prior appropriation, where the legal specification has a volumetric basis and contains an additional variable - seniority. Entitlements under prior appropriation are, however, hedged by ancilliary public interest obligations, some judge-made and some statutory. In addition to these common "state" variables the degree of attenuation of the private right is further extended by the presence of effluent obligations coupled with common law recognition of "injury". The reality of return flow obligations is evident in principle from the fact that an upstream entitlement can be junior to a downstream entitlement. Court supervised transfers of "native" entitlements under prior appropriation often determine a <u>transferable</u> entitlement which is a minor fraction of the volumetric face value or "paper" volumetric entitlement. This arises from judicial recognition of effluent claims, which can often amount to 60 to 80 per cent of the water right. Consequently, while transferability is at least possible under prior appropriation and while additivity is present in principle, the relationship between "paper" water and "physical" water is determined by implicit obligations that do not appear on the title. Attenuation of the property interest is a stark fact when the transferable element of a right is but a small fraction of the nominal right.

Under the American constitutional system "new" water supplied by Federal or other inter-basin transfer projects can be stripped of most of the implicit attenuating influences that apply to native water resident within its natural catchment. Depending on the local rules that apply, "new" water can be highly transferable at, or close to, face value. Celebrated cases of free transferability of non-attenuated titles, such as those supplied by the Northern Colorado Water Conservancy District from the Colorado-Big Thompson system depend on these attractive possibilities that are attached to "new" water.

The "capacity share" approach to specification of entitlements, if comprehensively applied, would call for an universal and exhaustive translation of paper rights into a new two-part entitlement. The first would be an unattenuated right to a volume less than or equal to the prior attenuated right. The second part would be a right of access to the residual of the "paper" entitlement on the basis of a 100 per cent effluent obligation.

Colorado and some other "prior appropriation" States have well-established arrangements for transfer of "native" water entitlements. Even in these instances, "retail" transfers are cumbersome and costly to achieve. It appears that major practical improvements in the efficiency of economic influences on water use can best be achieved by redefining entitlements of major systems operated by supply authorities. Retail markets almost always exist in the shadow of the big systems; the rules operating within such systems can be considered to be derivative of the rules applying between them.

X. HIERARCHY OF ENTITLEMENTS

To create the preconditions for application of efficient economic instruments in water resources management, a hierarchical structure of rights should be employed. Capacity shares would be defined to rationalise allocations to supply authorities with allocations to agencies with existing fisheries and environmental responsibilities, and those that use water for hydro and thermal power generation. "Retail" allocations to private end users - irrigators, stock and domestic users, groundwater licensees, major industries and the like, would continue to be based on existing "retail" instruments of entitlement. Measures could then be introduced to allow these subordinate rights to be progressively freed up so as to become substantially transferable, and hence, less attenuated than at present. The main benefit to end users from new "wholesale" instruments would lie in the scope which they provide for better definition of volume/security attributes of retail right. When the bulk supply authority has an exclusive claim over its share of the resource, it will be possible for that authority to provide a more sophisticated range of "products", tailored to the needs of particular classes of end users, subject to better guarantees of delivery, and with a less attenuated title.

Capacity sharers would be obliged to make their own decisions about security policy, and to take full responsibility for their own operating policies. If they chose to maximise yield at the expense of security, the consequences would fall upon the responsible authority, rather than being diffused across a number of users of a shared source. Bulk shareholders may choose to make their own hydrological forecasts, or to rely on expert forecasters. That is a purely technical and financial choice. They would, however, not be able to escape the obligation of making their own carryover decisions, and bearing the consequences. It is expected that supply authorities will listen to the opinions of their own consumers when making centralised choices on behalf of classes of consumers or "segments" of their market.

Supply authorities will be far better placed than they generally are at present to be responsive to the wishes of users, because the results will be exclusively in their own hands. The authority that bears extra cost in the interest of achieving greater security will not be in danger of having the benefit appropriated by others, who, having taken bigger risks, face the prospect of going short in a dry year.

From a technical viewpoint it would be feasible to extend the capacity sharing approach down to the "retail" end user. Each end user would take a portion of the bulk share owned by the supply authority, together with an "effluent" component. Dudley and Musgrave (1987 forthcoming) have explored some aspects of "retail" capacity sharing by means of simulation/optimisation techniques, using data from some New South Wales irrigation regions. The additional complexity introduced into retail entitlements by distribution losses, surplus flows, tributary inflows below the storage and transport time lags, cut into the potential benefits that are available in principle. Transaction costs in the creation of new titles and the additional information costs to the end-user are likely to be disproportionately high when associated with relatively small "retail" volumes of water. Improved transferability of water entitlements at retail level offers significant gains through reduced attenuation of the irrigation right, without incurring the additional cost and complexity of retail capacity sharing. Hence it is concluded that at the "retail" end of the supply system, the initial transaction costs and the ongoing information costs of capacity sharing would outweigh the benefits of capacity sharing.

XI. ENVIRONMENTAL FLOWS

Operating rules which formally recognise the protection of riverine conditions as a claim on regulated flows are likely to be part of the package in future water resource development projects in most developed countries. Construction of a dam or weir can completely intercept streamflows from upstream of the storage, radically altering fish habitat and conditions for riparian vegetation, together with salinity levels and other aspects of water chemistry in downstream wetlands, lakes and estuaries. Environmental quality is effected to some extent by any barrier to streamflow, but the effect can range from trivial to catastrophic, depending on the character of the stream and the extent of flow modification. Most existing legal frameworks do not lend themselves to formal provision for environmental flow rules.

A commitment of releases for environmental purposes will often draw on stored water, or at least on water entering the storage. Existing environmental commitments, where they exist, can usually be regarded as a form of release sharing. Like any other release share, they are underspecified relative to the equivalent capacity share, but like other release shares, they can be translated into capacity shares and hence can be given a non-attenuated form.

Without denying the difficulties that the specific physical properties of water systems present for creation of an efficient management framework, arcane features of public goods attributes of water have, in my opinion, been over-emphasised. Because of its stochastic character in both demand and supply, water presents some special management difficulties. These difficulties have, however, been gratuitously amplified by vagueness in the legal instruments of entitlement. Gratuitous complexity has been nowhere more damaging than in the assignment of the public/private goods distinction.

Of course there are conflicting demands for wild and scenic rivers and for regulated water sources, for fish habitat and for effluent disposal, for recreational lakes and for fluctuating storage levels. Happily it is for once possible to say that water is in no sense unique. Like virtually every other element of the human environment, water is subject to competing demands of private and public uses. It is primarily the measurement problem that has confused analysts and misdirected attention in public choice away from effective demand toward normative contention. The discipline involved in translating underspecified entitlements into capacity shares has the side benefit of translating many vexed environmental issues into clear questions of public choice, once the veil of gratuitous fuzziness is drawn aside.

XII. THE CONSTRUCTION IMPERATIVE AND RESOURCE RENTS

Whether the focus is on the United States Bureau of Reclamation and the Corps of Engineers, or ubiquitous urban supply authorities around the world, or the irrigation agencies of the Australian States, economic critics have regularly lamented certain common features of institutional behaviour. The construction authorities have characteristically enjoyed a popular mandate to invest in additional bulk supply facilities ahead of effective demand. Incremental costs have frequently been disguised from final consumers through subsidies from general taxation, or from cross subsidisation by other water users, or by means of tariff structures more or less unrelated to the individual consumer's consumption choices. Costs brought to account by supply authorities have typically included part or all of facility costs, but raw water sources have been accounted as costless. In maintaining an artificial surplus of supply over demand by excess investment, supply authorities have commonly maintained resource rents at or close to zero.

During the past century or so a special set of political circumstances and consequential bureaucratic imperatives, together with freedom to graze on the public purse, drove on the many bodies with mighty source development powers. These circumstances appear to be on the way out in many countries. The age of the dinosaurs is not yet over, but it does appear that in many parts of the developed world they are in decline under ubiquitous attack from smaller smarter species. Their demise or adaptation to the new conditions for institutional survival is a proper cause for celebration, though it may perhaps be tinged with nostalgia in some quarters. In many parts of the world, economic rationalism is rapidly overpowering the primitive drive to build in advance of effective demand. This evolutionary transition bears a promise that substantial institutional imperfections will gradually be eliminated. Demand will more directly trigger new supply and the reallocation of existing supplies. Economic rents will emerge so that the natural water resource will no longer be a free input to the production of water services. Such a transition represents, in itself, the emergence of a new pattern of efficient economic incentives in water resource management.

The statutory creation of a new and un-attenuated instrument of entitlement can play a major part in the transition from a supply-driven to a demand-driven water management system. If a non-attenuated entitlement can be defined, each new regulated source of supply can be exhaustively partitioned in advance amongst specific beneficiaries. This in turn can permit a precise match between shares of cost and shares of product. Vague specification of future entitlements has often facilitated the pea and thimble trick that was for decades the key stock in trade of the builders when obtaining project authorisation.

XIII. OBSTACLES TO TRANSITION

It appears that proper definition of water resource bulk entitlements in terms of capacity shares offers a basis for water resource management that is clearly superior to existing arrangements. To justify a change there should be sufficient latent benefits available from a capacity sharing regime to compensate the losers and still be in position to realise net gains. Time is a great solvent; the convenient occurrence of drought sequences or environmental disasters can present political "windows of opportunity" when constituencies emerge that seek new arrangements. Often it can be demonstrated at times of crisis that much of the political heat is caused by disagreement as to factual or even accounting matters.

In Victoria, Australia we hope to obtain enabling legislation which permits the old and debased coinage of release shares to be translated into the new currency of transferable capacity shares. Statutory safeguards and saving provisions are proposed to confirm the negotiating base of existing right holders. The rate of uptake of the new instrument of allocation can only be guessed at. There are currently enough supply authorities with exposed positions to offer a high probability that some immediate use will be made of the new provisions. (Dept of Water Resources 1987a, 1987b).

XIV. MARKET SYSTEMS ARE REGULATION BY OTHER MEANS

The fairly recent upsurge of professional interest in the use of markets for natural resource management has coincided in the United States, Britain, Australia and many other countries with the emergence of "neo conservative" and "new right" ideologies. Just at the time that regulatory intervention has been recognised as only one part of the efficient response to market failure, markets have simultaneously been misperceived by liberals as an adjunct of the conservative agenda.

In the natural resources management field, appropriate application of markets calls for political choices that are not excessively coloured either by the neo conservative prejudice that markets can fix everything or by the liberal and environmentalist concern that markets are the stalking horse of political reaction. Economists have discovered the importance of property rights too recently for more than a few members of the profession to have thought seriously about the positioning of rights in the spectrum of allocative and transactional devices.

Property rights have never been absolute in any political system. At minimum they have conferred a right to recovery in the courts, but the means of recovery are rigorously conditioned by law. Real property conferred a right of action against trespass, but rarely conferred an absolute right to murder the trespasser. Common law doctrine of nuisance and waste-confined use of property to "quiet enjoyment" and to customary forms of stewardship. Some but not all western societies prohibit "taking" or appropriation by governments but this merely creates a claim for just compensation; it is no barrier to compulsory acquisition. Powers are usually available under common law or statute to allow public authorities to take easements. Real property rights may or may not give title to minerals and groundwater. While doctrines vary, drainage rights, and obligations to receive drainage from others limit development entitlements. The broader legal framework and police powers underpin the "enforceability" of property rights, but the extent of these rights is a pure artifact of the legal, and ultimately the political system.

For all these reasons it can be said that every property right is severely attenuated. That is, in itself, no barrier to the existence of efficient markets in titles, provided that all the legal limitations on title are uniformly and impersonally applied, and provided that expectations are stable. Every economist is aware that the theoretical conditions for the existence of a perfect market are never satisfied in reality. That recognition has neither nullified the potential of markets in many practical fields of resource allocation, nor deterred the profession from making analytical use of the market concepts. Property law is less familiar territory to most economists but very similar considerations apply; the use of appropriate concepts of proprietory interest can offer analytical and empirical power, even though proprietory interest is always highly conditional.

No title is more conditional than the title to urban land. The common law conditions mentioned above are all applicable except where they have been replaced by statutory counterparts. In addition, the "native" limitations are supplemented in most cities with highly detailed and specific land-use planning codes and statutes, designed primarily to control externalities and to enhance local security of expectation. Within all these constraints urban real estate markets still perform as highly efficient instruments of allocation.

Thus it is quite incorrect to conclude, as some appear to do (Anderson 1983) that attention to proprietory interest will, of itself, advance the neo-conservative cause. Titles and marketable instruments are technical devices which, properly employed, can enhance the efficiency of almost any political agenda in a liberal society. Attention to the definition of rights offers enormous opportunities for enhancement of the technical efficiency of water allocation decisions. Better definition of rights is a precondition to creation of more efficient allocative frameworks, but property rights and efficient markets are mere technical servants of broader social purposes; they are not the key to any neo-conservative Nirvana. Politics and administration, as well as the courts, will continue to fix the underlying parameters that condition market outcomes.

REFERENCES

1. Alaouze, C. M. and Tai, P., "A Market Solution to the Bulk Allocation Problem Under Release Sharing" (Appendix IX, DWR 1987b)

2. Alaouze, C. M., "Reservoir Yields Associated with a Specific Partitioning of Reservoir Releases to End Uses Characterised by Different Reliability Criteria", (Appendix VIII, DWR 1987b)

3. Buras, N. (1985) "An Application of Mathematical Programming in Planning Surface Water Storage", Water Resources Bulletin, 21 pp.1013-20.

4. Anderson, T. L. (ed) (1983) Water Rights, Scarce Resource Allocation, Bureaucracy and the Environment, Ballinger Publishing Co., San Fransico.

5. Cheung, S. (1970). The Structure of a Contract and the Theory of Non Exclusive Resources, Journal of Law and Economics, Vol. 13, pp. 49-70

6. Department of Water Resources (DWR) (1986a). Water Law Review Discussion Paper, (Water Resource Management Report Series, Report No. 1, VGPO, Melbourne).

7. DWR (1986b) Water Law Review : Transferable Water Entitlement (Issues Paper Water Resource Management Report Series, Report No. 2, VGPO, Melbourne).

8. DWR (1986c) Water Law Review: (3) Water Law and the Individual: (4) Riparian Rights: (5) Dam Safety (Issue Papers Water Resource Management Report Series, VGPO, Melbourne).

9. DWR (1987a) Executive Summary; Flexibility in Major Water Allocations (VGPO)

10. DWR (1987b) Flexibility in Major Water Allocations (Background Paper VGPO)

11. Dudley, N.J. and Musgrave W.F. (1986) Property Rights and Capacity Sharing of Surface Water Resources. Contributed Paper, 30th Annual Conference of the Australian Agricultural Economics Society

12. Dudley, N.J. and Musgrave, W.F. (1987). Capacity Sharing of Water Resources (forthcoming).

13.   Loucks, D.P., Stedinger J.R. and Haith (1981) Water Resource Systems Planning and Analysis, (Prentice-Hall, New Jersey)

14.   Paterson, J., (1986) "Co-ordination in Government: Decomposition and Bounded Rationality as a Framework for User Friendly Statute Law", Australian Journal of Public Administration, Vol. XLV. No 2 pp. 95-111.

14.   Peters, T. J. and R., H Waterman Jr. (1982) In Search of Excellence: Lessons from America's Best Run Companies, (Harper and Row)

15.   Randall, A. (1975). "Property Rights and Social Micro-economics", Natural Resources Journal, Vol. 15, pp. 729-38.

16.   Randall, A. (1978). "Property Institutions and Economic Behaviour", Journal of Economics Issues, Vol. 12, pp. 1-21

17.   Randall, A. (1981). Resource Economics : an Economic Approach to Natural Resources and Environmental Policy, (Grid Press, Columbus, Ohio)

18.   Simon, Herbert A. (1983) Reason in Human Affairs, (Stanford U.P.).

Chapter 4

WATER RESOURCES MANAGEMENT AND THE ENVIRONMENT
ECONOMIC INCENTIVES IN THE NETHERLANDS

Dr. Paulus J. Baan

I. INTRODUCTION

This paper focuses on water resources management policies in the Netherlands. Some of the basic principles in planning and decision-making will be discussed as well as the experiences with the economic incentive mechanism.

Unlike many Anglo-Saxon countries, the Netherlands does not have a system of water rights (be it to riparian doctrine or on the basis of prior appropriation), so that the Dutch water management setting is very distinct from the conditions described by Frederick (Chapter 2). In his paper, Frederick emphasizes that water rights, if transferable and properly applied, may constitute a very efficient means to achieve optimal use of water resources, citing the example of the Westland Water District in Southern California.

In the Netherlands formal water rights have never been applied, nor have government or others ever proposed to do so. The reason for this can be found in the history of Dutch water resources management. Being a delta area, protection against floods and reclamation of land were historically the first tasks in water management. Over the years this has resulted in a vast system of dikes, sluices, locks, weirs and pumping stations that now allows nearly full control of surface water even under extremely wet or dry conditions. These tasks could only be performed through collective action for which the direct beneficiaries (mostly the activities and residents of the respective polder) were financially responsible. Management of the "assets" was put in the hands of so-called water boards of which the representatives were elected among the local people. This form of government (functional government) actually preceded any other form of modern government in the Netherlands. It is clear that in those early days water could not really be considered a resource posing problems from the viewpoint of conflict of interests, unlike present day society. The institutions which developed out of such conditions, in which every waterboard could handle its problems separately (as if it were in its own catchment, cf. Paterson [Chapter 3)], nowadays must deal with problems requiring different organisational, legal and financial arrangements to better respond to the new conditions. Before addressing the role of economic incentives in the allocation of resources, first the nature of the use of the present day water resources system in the Netherlands will be presented in brief to serve as a background for the rest of the paper.

II. NATURE AND USE OF THE DUTCH WATER RESOURCES SYSTEM

With respect to water resources a distinction has to be made on the one hand between surface water and groundwater and on the other hand between quantity and quality. In the Netherlands surface water can generally be withdrawn easily at low cost. No charges are levied, nor are water rights involved (see above). Most of the investment costs and all operation and maintenance costs for the supply of this type of water have to be fully covered through taxes imposed by the waterboard on economic activities and on the residents of the waterboard area. The waterboard tax also covers costs for dike maintenance and drainage improvement. The water supply component of the waterboard tax normally is relatively small compared to these other components.

After use, the water can be returned to the same surface water source, to another water body, or it can be re-used elsewhere. Discharge of used water is subject to an effluent charge, imposed by the Pollution of Surface Waters Act of 1970. The charge is based on standardized "resident equivalents" (r.e.'s) and is used to recover the total costs of waste water treatment facilities, including subsidies provided to private treatment facilities. Among the most important issues in surface water management are the impacts of consumptive use with respect to quantity management, especially in dry periods, and possible changes in the surface water quality of the local and national water systems as a result of the intake of water from other water systems, the discharge of waste water, and instream uses such as inland navigation.

For households and industry average consumptive use is about 10 per cent. The same percentage holds for public water supply as households and industry are the main users. Agricultural use, on the contrary, is highly consumptive. In finding an optimal use pattern of water the possibilities of re-use have to be accounted for.

Using the elaborate water management system described above, surface water can be transported to most parts of the country. Costs of transport, however, may make this economically unattractive. The same holds for quality. Good quality water, which is freely available at many places, is a valuable resource. Low quality water can be transformed into high quality water at certain costs. These costs may be high compared to a free source of good quality water elsewhere, thus making bad quality water an economically unattractive source. Consequently, upstream consumptive water use and water pollution can place a heavy burden on downstream users.

Abstraction of groundwater in the Netherlands is generally more expensive than withdrawal of surface water. In most cases, however, groundwater is of superior quality and therefore preferred as a source of raw water. After use, groundwater is usually discharged to surface waters (groundwater recharge is an exception in the Netherlands). Therefore, all groundwater abstractions can be considered as consumptive use. Shortages of groundwater (dropping groundwater levels) lead to competition between present and potential users. In this respect, concern has arisen recently about water in the context of nature conservation to the extent that surface waters may be seriously affected by excessive groundwater use.

Groundwater extractions in excess of about 10 m3/hour are subject to a license system. If extractions are proven to be harmful to other parties the applicant has to compensate the damaged parties for the total capitalized financial damage.

Groundwater serves as the major source for public water supply, but in the western part of the country demand exceeds supply and surface water has to be used to a large extent. Treatment costs for surface water are high compared to the costs of groundwater treatment. As a result the costs of public water supply and the rates charged show large differentials among the various regions.

The quality of the groundwater aquifer is not affected by the withdrawals. Groundwater pollution mainly stems from land-based activities like farming and therefore quality control problems are quite different from quantity management problems.

The time dimension of the effects of pollution is an important factor in water management. When lakes, canals and rivers are polluted, cleaning up the sources of pollution will result in a relatively quick quality recovery due to the 'flushing' by good quality water. The recovery process is retarded, however, by polluted sediments. Pollutants from such sediments may be released slowly, affecting the quality of the water for some time. This retardation becomes more important at low flushing rates, resulting in greater residence times of water, as in lakes. Even then, recovery times are still much shorter than for groundwater. If groundwater is polluted, it may take hundreds of years before the quality is restored. Such aquifers can no longer be characterized as renewable.

From the preceding it can be seen that economic incentives to use and develop existing resources efficiently are not explicitly used in the Netherlands. On the contrary, Dutch policy-makers have always rejected the predetermined use of economic incentives in water resources management: optimal water resource use is pursued mainly by non-market policies such as agreements or licensing. In consultation with water users and 'representatives' of other water-related interests an optimal use pattern is determined and policies are adopted to achieve this pattern. But how does planning in such a system of resource allocation deal with economic principles? The section below presents a study which has had a major impact on the planning process at the national level, the methodology for which was later introduced at the regional planning level.

III. PLANNING OF WATER RESOURCES

From 1976 to 1980 a large study ("PAWN": Policy Analysis of Water Management in the Netherlands) was undertaken to develop alternative water management policies for the Netherlands and to assess and compare their consequences (Goeller, 1983). The PAWN study aimed at finding an optimal allocation for fresh groundwater resources (in view of growing demands) and of surface water. The supply of surface water is sufficient except in dry years, when there is competition not only among water-use categories as agriculture, power plants, shipping, and nature, but also among different regions. In dry periods the basic trade-off in the Dutch national system is between water supply to the northern and eastern higher regions (to prevent agricultural damage from drought) and the water needed in the western region to prevent salt intrusion, which may damage the crops in high-value horticultural areas.

The study was primarily supply oriented: demands for water by traditional economic activities were assumed and possibilities to reduce demand were analyzed only with respect to industrial water demand. Relatively little attention was paid to water quality. Only salinity, eutrophication, and thermal pollution were studied. Based on the results of this study a policy document on national water management was written by the Ministry of Transport and Public Works. In this document it was declared, inter alia, that the existing national physical infrastructure was adequate for the next 10-20 years and that groundwater should be reserved for public water supply and for high value added uses in industry. Withdrawals for the purpose of watering agricultural crops have been exempted, because of the expected administrative problems and costs associated with having to issue and control numerous licenses for relatively minor withdrawals.

In the PAWN study economic criteria played a very important role in determining desirable water management strategies. For example, in screening existing or anticipated local and regional water management plans which were designed to increase or ameliorate water supply to agricultural areas, benefit-cost analysis was used. Expectations of future demand in agriculture were based on benefit-cost calculations of farm budgets with and without such plans. The fact that economic criteria were explicitly used in this study however does not guarantee optimal allocation of water resources. Some of the shortcomings of the economic approach used in this study and in the actual daily operation of the system will be discussed in section 5.

Since 1980, insight into water resources management has grown. Now emphasis is placed on the functions related to water resources. This is done by following a so-called water systems approach as described in the policy document "Living with Water" published recently by the Ministry of Transport and Public Works. The approach aims at optimal coordination of the preferences of society with regard to the functions and the functioning of the water systems. A water system in this respect is defined as the geographically demarcated, interrelated and functioning whole of surface waters, groundwater, underwater beds, banks and technical infrastructure, including the existing eco-systems and all related physical, chemical and biological features and processes. The boundaries of a water system of this kind are determined in the first instance on the grounds of morphological, ecological and functional relationships.

As a result of extensive planning and interaction among all interests, water resources management in the Netherlands may well yield results close to the socially-desired optimum situation. However, problems still exist, especially with respect to water quality. As a result of a massive clean-up operation of waste water discharges in the early 1970s, now nearing completion, water quality problems nowadays more often stem from interaction of water systems with the other environmental media (pollution of air and soil). Thus, in water resources management, primarily non- water related activities like applying (excess) manure in agriculture are getting more and more attention. Due to run-off, applying excess manure leads to diffuse pollution of surface water, and as a result of leaching, e.g. of nitrate, groundwater is polluted. This kind of 'inter-environmental' relationship and the economic activities concerned are the major problems for which solutions have to be found.

IV. INSTITUTIONAL ASPECTS

Most present day water resources planning is based on functional classifications. For surface water a distinction can be made between instream and withdrawal uses (cf. Feenberg's benefit categories 1980). Instream uses encompass recreation, nature conservation, fisheries and shipping. Shipping is dependent on the quantity of water (depth of rivers and channels), the other instream uses are dependent mainly on the quality. Industry, households and agriculture are typical withdrawal uses for which both quantity and quality are important. In searching for an optimal use pattern all uses have to be taken into account. Costs and benefits for withdrawal uses and shipping are assessed relatively easily and, as they are expressed in monetary terms, they can be understood quite well. With respect to the other functions (especially recreation and nature conservation) valuation of costs and benefits is more difficult. As a consequence and due to the fact that withdrawal uses and shipping represent economic sectors which are market-oriented and which are represented through well-established organisations in the decision-making arena, these latter functions often get more attention than recreation and nature conservation, and can lead to a distortion of the optimal use pattern. Due to political pressure from nature conservationists since the 1970s this is changing.

With respect to groundwater it is more or less the same. The withdrawal functions (households, industry and agriculture) usually get more attention than the other functions, which may be described as on site functions (recreation and nature conservation).

As stated before, upstream water use affects the downstream water use possibilities. For example, salt discharged in the river Rhine by French potassium mines causes damage to horticulture in the western part of the

Netherlands. Therefore in seeking an optimal use pattern for water the whole basin should be considered. With respect to the river Rhine this means that country representatives should meet and agree upon water use from the river basin and on the way how to realize an optimal (international) water use pattern.

The issue of international water rights (also covering the water quality aspects) or licensing practices should also be examined. The dumping of salt in the river Rhine is illustrative in this context. This case was brought before a court, which pronounced that both parties, as users of river Rhine river water, had to mutually consider each other's interests. The 'polluter pays' principle was not strictly applied, as is often the case where economic interests are high.

Financial arrangements pose many problems for effectively exercising integrated water management. One of the most common ways in the Netherlands to improve the surface water quality situation in a water system is through flushing. Such measures, however, cannot be financed, because quantitative surface water management is financed out of waterboard taxes or by the national government, and surface water quality management is financed solely through effluent charges. According to the 1970 Act, the effluent charge does not cover quantitative water management measures. In the Act, it was not foreseen that water quality management might also imply measures to improve water quality in a system for functions such as nature and recreation. Hence, at the moment, there is no financial basis for such measures. Furthermore, quantitative water management is controlled by other institutions than water quality management, which introduces the need for intensive interactions, often hampered however by the lack of a sound financial basis.

V. ECONOMIC ASPECTS

In the Netherlands average pricing is applied to public water supply instead of marginal pricing, which would result in more optimal allocation from an economic point of view. Moreover, these average prices are based on historical costs, which means that aged water works charge low prices far below average replacement costs. Also, large users of public water like some industries pay less for water, given the low block rates charged for larger quantities. In such cases, the costs to industry are lower than the marginal costs of producing public water. This favours (too) high water demand. Average historical cost-pricing is in fact prescribed by the government, and is thus perhaps more representative of a "government failure" than a "market failure".

Because of the low costs of public water supply, households and industry are hardly conscious of water prices. Though prices for public water in the Netherlands differ by region by about a factor of two, no clear differences in demand by households are observed. Therefore the price elasticity of demand must be small. Energy prices are more important since energy accounts for a larger part of the household budget. Part of the water demand (e.g. for washing and bathing) is related to energy demand and therefore water demand may decline with higher energy prices.

It is interesting to note that water demand by industry diminished in the last decade. The two main causes were the imposition of (rising) effluent charges and higher energy prices, the latter resulting in energy savings and a corresponding reduction of cooling water demand. The increasing effluent charges provided an incentive to reduce waste water discharges and increase recirculation of water, which in turn further reduced water demand.

In agriculture, horticulture and cattle-breeding, many products are subsidised in one way or another, either directly or indirectly. Horticulture is subsidised indirectly by low energy prices. For many agricultural and dairy products guaranteed minimum prices are set by government. This occurs also in the European Community context. As a consequence overproduction is ubiquitous, entailing high storage costs. From an economic point of view this overproduction can be considered as leading to losses for society and allowing water to be withdrawn for this purpose can be considered a misuse of valuable resources.

These market distortions may play an important role in the way planning studies are carried out. As mentioned before, in the PAWN study cost-benefit analysis was applied (Goeller, 1983). The benefits of dairy production were based on the subsidised market prices, thus leading to an overestimation of the value of water for dairy production.

Another distortion was introduced through the use of models which sought to simulate farmers' behaviour. Calculating water demands according to a formal farm budget analysis which included, amongst others, payment for the labour of the farmer-owner, did not correctly predict the water demand pattern (especially for sprinkling). Farmers were sprinkling more than the budget analysis indicated. Apparently farmers valued potential crop losses higher than the cost of their (labour) inputs. It is clear that in determining real agricultural water demand, market prices and actual water user behaviour should be used to calibrate the models.

In the final elaboration of economic strategies on the national level, however, water should be valued at shadow prices in order to correctly reflect the scarcity value of water allocated to farmers. For various reasons such 'corrections' were not applied throughout the whole study, although in some instances results of 'corrected' calculations were presented next to the results of calculations based on market values. The omission of these corrections in the PAWN study did not influence the major conclusions of the study: even with distorted values for agricultural water use none of the possible major infrastructural measures proved to be cost effective. This, however, is not the case with regional water management plans, so that at this time economically inefficient allocations and investments may still be made.

Subsidisation in the cattlebreeding sector also results in large quantities of (excess) manure leading to surface water pollution (run-off) and groundwater pollution (leaching of nitrate). Pollution of groundwater in particular is a long-term problem. Some public water companies in the Netherlands are already facing high nitrate concentrations in their groundwater source and have had to take counter measures. It is expected

71

that these problems will worsen over the next ten to thirty years. If excess manuring is continued in this way, groundwater sources will be polluted for many hundreds of years. Farmers essentially do not use groundwater extensively as a source in a quantitative sense. In a qualitative way, however, groundwater is used for dumping of waste products, of which excess manure is a component. Therefore farmers should be considered users of groundwater resources and should be taken into account in finding an optimal use pattern. A complicating factor is that pollution of groundwater (with nitrate) is an irreversible process turning groundwater into a non-renewable resource for a very long time.

A more theoretical remark has to be made with respect to the time factor. Costs and benefits that emerge over time are normally discounted down to net present values. As a result long-term effects are valued at a very low level. Problems arise in cases where the natural environment might be affected and where these effects turn out to have irreversible impacts on the quality of the environment and of life. According to Hueting (1980) discounting involves the risk of underestimating the value of natural resources by the present generation. In his opinion, for long-term effects, the preferences of the current generation should indeed not be used to value these resources. In light of the above, the following questions arise:

-- How can economics deal with possible long-term effects which in one way or another might be essential for the quality of environment and life and which are neglected as a result of discounting?

-- How would future generations value foregone benefits (e.g. of the use of certain resources) and by whom are future generations represented?

It has to be remembered that the optimal use of water resources in the short run may differ from that for the long run. Economic and technological development are important factors in this respect. Long-term investments made to overcome inadequate water supply conditions or bad quality conditions may be important too. After such investments have been made, users normally become less concerned with water as an important scarce resource and consequently the availability of good quality water is valued lower. In the long run, however, when reinvestment is considered, water again is valued higher.

Two examples of such investments in the 1970s in the Netherlands are worth mentioning. In the southwest part of the country a large reservoir was built to tide over periods of bad water quality in the river Meuse, which is the major source of public water supply in that area. In the same period in the southeast of the country, two large cooling towers were built as stand-by for a power plant to tide over in case water supply from the river Meuse would be short. Once these problems were countered, politicians and the general public became relatively less concerned with possible future problems of the same kind.

VI. TOWARDS IMPROVED POLICIES

Society becomes more and more complex. Owing to this complexity difficulties in finding an optimal water use pattern grow. Water use (both in a quantitative and a qualitative sense) is dependent on and interrelated to many factors. The value of water for different users cannot always be established easily. Markets do not operate well since nature conservation is not conducted on market principles. Dealing with the conservation function in isolation entails the risk of its under-evaluation, especially in the long run. Therefore an important role is reserved for government in its ongoing search for the optimal solution. Integrated studies are needed to identify these solutions and to determine effective policies for attaining these goals.

The need for further study can be illustrated by the experience of industsries with waste water management. The design of many municipal waste water treatment plants is based on the number of households discharging to the sewerage system and the industrial waste load at that time to the same system. The costs of treatment had to be allocated to households and industries on the basis of their waste load discharged. Faced with these costs, industries started to search for ways to reduce their waste load and many succeeded, for example, by greater recycling. As a result the total waste load decreased and charges per unit of waste load went up, motivitating industries to yet further reduce their waste loads. In the end, treatment plants were confronted with excess capacity, for which society had to bear the costs. A preliminary study demonstrating the incentive character of this particular financial approach could have prevented this.

An important issue in studying water demand and optimal water use is the relation of water demand to the costs of other resources and raw materials. In this respect, the case of increasing energy prices and the effects thereof on water demand was mentioned. Of similar nature is the problem of effluent charges. High charges on waste water give industries incentives to avoid them by increasing recirculation of water, resulting in decreasing water demand. The financial impact of these effects should be properly accounted for in developing effective and economically efficient water quality management strategies.

Another example is agriculture. Agricultural production is subsidised and minimum prices are set for a number of products. Over the years concern about the high costs of excess production has grown. Measures like charges on excess dairy production and manure generation are now being taken to decrease production. As a result the value of water in agriculture is declining to a point where politicians can no longer close their eyes to indirect subsidisation of this sector through the supply of water at a price which may well cover the total costs to the water supplier, but not the costs to society. There are already indications that the change in market conditions has had an impact on the willingness of farmers to pay for the present system of water management. Farmers are opposing new water board plans to increase the supply of fresh water but which at the same time will increase the water board tax to levels unacceptable to them.

In the policy document 'Living with Water' the financial problem was explicitly raised. New financial arrangements will have to be developed to provide the necessary means to exercise active integrated water management. This probably requires a new set of water management and environmental protection laws, because the existing laws were simply not designed for this purpose. Environmental protection strategies no longer strictly apply the 'polluter pays' principle. More and more, government is seeking to dialogue with industry and other polluting sectors to jointly find ways to reduce waste disposal in the environment. This policy is prefered over economic incentives and market-oriented actions, because it creates a basis for an open dialogue in which more information is transferred between the actors than is possible through one-sided instruments like charges and taxes. It is hoped that this fuller flow ultimately results in solutions suitable to all parties involved.

REFERENCES

1. FEENBERG, D., et al. (1980), Measuring the benefits of water pollution abatement, Academic Press, New York.

2. GOELLER, B.F., et al. (1983), Policy Analysis of Water Management for the Netherlands, Vol. I, Summary Report, Rand report R-2500/1-NETH, Santa Monica, March 1983.

3. HUETING, R. (1980), New scarcity and economic growth, North-Holland Publishing Company, Amsterdam.

4. Ministry of Transport and Public Works (1985), Living with Water: Towards an Integral Policy, The Hague.

Chapter 5

FOREST RESOURCE MANAGEMENT AND THE ENVIRONMENT:
THE ROLE OF ECONOMIC INCENTIVES

Dr. Roger Sedjo

I. INTRODUCTION

The forest produces a variety of commodities and environmental services. The lack of clear property rights to the forest and/or some of its several outputs, together with the lack of well-developed markets for some of its products, has created periodic common property and externality problems.

This paper examines the nature of economic incentives and the sources of inefficiency in the management of the forestry sector with particular focus on the environmental consequences. The essential features of economic incentives as applied to forestry management and their impact on efficiency and environmental systems are investigated. Inefficiencies of two general types -- market failures and government failures -- are examined. Market failures usually occur when property rights are poorly defined or when the distribution of property rights leads to the inability of externalities to be internalised. Government failures refer to inefficient resource management either directly by government agents as managers of a publicly-owned resource or via regulatory interventions which lead to inefficient management. A major focus of this paper is the consequences of these two types of failures on the environment.

In response to market failures, two approaches to governmental intervention have emerged: 1) regulation and 2) public ownership and management. However, in the United States, the management of the forest under both of these regimes has come under heavy criticism by economists as well as by special interest groups. Governmental failures have been alleged both in the regulatory process and in public management.

These issues are examined within a broad historical context tracing the evolution of the forests and of the institutions developed to manage them. While the focus is largely on United States' experience, global experience will also be cited. The strengths and weaknesses of the various institutional management systems will be discussed, with particular attention given to the tendencies to achieve efficient resource utilisation via the incentive structures associated with each institutional setting.

Critics of public forestry management in the United States have suggested two general types of basic remedial changes. The first would reform the existing public management system through improved legislation and techniques such as forest planning models. The second involves large scale restructuring of control and management functions to render decision-making more responsive to market signals and incentives. This second type of change covers the spectrum from changing the signals to which public agencies are required to respond, to privatisation or quasi-privatisation of public timberlands. Management decisions would thus be more freely responsive to market signals and incentives. These proposals are comparable to schemes which seek to introduce market-oriented regulatory mechanisms such as the sale or transfer of water or pollution rights as a means of achieving increased economic efficiency.

II. FORESTS' ROLE IN HUMAN HISTORY

Forests in Transition

Forest resources have played a role in human development since the earliest times. They were a legacy of natural processes and were available to humankind on a "first come, first served" basis. In their natural state, forests provided a variety of commodity outputs and environmental services to early humankind. In addition to providing fuelwood and building materials, forests provided habitat for a wide array of plant and animal life that were used by humans for food and fiber as well as environmental services. When human pressure on forests was light, human management of the resource was not required and investments in regeneration and forest management were unnecessary. The natural resiliency of the eco-system was sufficient to insure natural regeneration and automatic renewal.

However, as human activities resulted in increased pressures on the forest, the disturbances to the forest system began to exceed its inherent recuperative capacity and deterioration ensued. In some cases humans began to actively intervene to "manage" the forests. Records report forest tree planting activities in ancient China and forest stewardship was important in Europe in the Middle ages. In more recent times, Europeans have gradually transformed the care and stewardship of the forest into the "science of forestry and forest management".

In many respects, forestry has had a history not unlike agriculture. Human food needs, initially met by a hunting and gathering activity, gradually came to be met by the agricultural activities and cropping and livestock raising. Similarly, humans initially met their need for wood for fuel and building purposes by foraging and gathering natural stands of trees. Today, we are in the midst of a transition in which the role of wood harvested from natural stands is declining as the world's industrial wood needs are increasingly being met through tree-farming activities. Just as agricultural production requires management and investment, so too wood produced by tree farming requires an environment in which economic incentives encourage good management and investments can take place.

But what of the nonwood outputs of the forest? While markets exist for industrial wood, markets are typically absent for watershed protection, wildlife and other environmental services provided by the forest. Under these conditions we cannot expect market principles to govern the provision of environmental services. Fortunately, in many cases environmental services are the byproduct of industrial forestry. This question will be addressed in greater detail below.

The Common Property Problem

Throughout history, forests were important assets providing a flow of goods and services to the surrounding population. However, as noted, little thought was given to protection or regeneration since this process occurred naturally and pressures on the resource were generally light. The earlier inherited forests in many parts of the world were, in practice, open access common property resources, i.e. they were the property of no one and hence, the property of all. Forest dwellers and others nearby could use the various outputs of the forest on a first come, first served basis. Approval was not required nor was reimbursement for the services of the forest. So long as demand for the various outputs of the forest was light, this system was adequate; natural regeneration was sufficient to replace the disruptions caused by humans. Conscious human management or investments in the forest was unnecessary.

However, as pressure built up on the forest the natural processes were unable to fully restore the forest unaided. The common property aspects of the forest also mitigated against management and protection. Since individuals had no claim to the fruits of management, investments in management and protection were not forthcoming. Property rights for the forest resource were missing. The absence of property rights led to a first come, first served behaviour pattern. If one person let a tree stand with the intention of harvesting it at a later more auspicious time, another person could fell the tree in the interim. There was no incentive to preserve, maintain and reinvest. This situation resulted in the classical common-property problem or "tragedy of the commons" (Hardin 1968). Over exploitation and subsequent degradation of the forest and the surrounding environment occurred due to a lack of clear responsibility and the inability to capture the returns to investments in management and protection of the forest.

Institutional Responses to the Common Property Problem

While degradation of the forest was a common outcome, it was by no means inevitable. Over time, various institutions developed which provided incentives for the care, protection and management of the forest. In traditional societies "informal" communal systems evolved with woodlands viewed as the property of individual villages, e.g. Japan (Osako 1983) or Nepal (Chapagain 1985). As more complex social systems emerged, this role came to be played also by the feudal landlords, centralised government or private property ownership. In the Middle Ages in Europe, the various fiefdoms commonly had rules for protecting and regenerating the forests. In the case of ancient China, the government took responsibility for protecting and investing in forests to ensure timber output as early as the 6th century B.C. (Menzies 1985). As the institution of private property developed, landowners had a strong self-interest to take the responsibility of the management and protection of forested areas under their ownership. After all, the owners had an interest in the value of their forestlands as income-generating assets, and well cared-for forest lands were viewed by the market as valuable, especially when occupied by valuable stands of timber.

From time to time however, these systems have broken down. While villages often protected common property forests in traditional settings, as the traditional societies dissolves, their forest protective systems also broke down and resource degradation ensued (Osako 1983). Systems of central government ownership also often had serious difficulties protecting the forest because the government lacked the ability and/or the will to do so. Simply placing the forest under nominal government ownership without the capacity to manage and protect, only recreates an open access setting that invites a "tragedy of the commons" outcome. (1) Private ownership too, often afforded inadequate protection especially when property rights were insecure or governmental institutions designed to enforce ownership rights were weak or absent.

In today's industrial world two major types of institutional settings have emerged which provide mechanisms to protect and care for the forests. The first of these is government control via ownership and/or regulation. The second is the institution of private ownership and private management decision-making. Between these are a variety of "mixes" of public and private ownership typically within the context of regulation.

III. MARKET FAILURES: RATIONALE FOR GOVERNMENTAL INTERVENTIONS

Market economists justify the heavy reliance upon markets by arguing that the combination of information and incentives provided by a properly functioning market promotes individual economic agents to actions that are efficient and in the broad social interest. Public intervention in private land use, including forestry, in a market economy is usually justified in terms of market failure. In the absence of a market failure, the private owner will have a set of economic incentives to manage the land to generate a level and mix of outputs through time. The owner will also have an economic incentive to undertake appropriate practices to maintain the fertility and productivity of the land. Should the land or forest quality deteriorate, the owner will experience financial loss as the market value of the asset erodes and will have incentive to invest in the restoration of the asset to the point where the marginal investment costs equal the marginal increases in the asset's value.

However, it is universally acknowledged that market failures do occur, often with natural and environmental resources. Market failures refer to situations where the signals (prices and costs) faced by the buyers and sellers do not adequately recognise externalities or third party effects. In these circumstances, signals provide incorrect information and correspondingly inappropriate incentives from the overall social point of view. Externalities are almost always associated with the lack of clear property rights and often can be overcome by redefining or reallocating property rights (Coase 1960). For example, the externalities associated with increased downstream flooding associated with increased logging, can, in principle, be internalised by either expanding the size of the private holding to incorporate the areas likely to be flooded or by negotiating an agreement between the logger and the downstream landowner to provide for payment for the damage of flooding (or a payment to alter the harvest to avoid the flooding).

In the absence of an internalisation of external effects, however, one would generally expect the rate of utilisation of a resource to exceed the socially optimal rate. In the context of a commonly owned forest, for example, the forest may be overlogged, experience erosion and delayed regeneration, thus affecting future timber yields. This problem can be readily addressed by conferring property rights to the forest. However, this does not always eliminate such problems for, as noted above, even if the forest is privately owned, ownership rights are typically not present for watershed and erosion control values generated by the forest. Thus, even if the market correctly determines the optimal logging rate for that site, a market failure could occur in the form of environmentally destructive harvesting practices that create downstream externalities such as flooding and silting. This problem is further complicated by the jointness of production of many outputs. For example, monocultural tree planting will provide some complementary environmental services, but it may also conflict with wildlife diversity. Clearly, given a traditional market situation and traditional assignment of property rights, the private owner has no economic incentive to manage his forest for its downstream/external environmental values. It is this failure that provides the traditional rationale for government interventions in the market process.

In contrast to private management of the forest responding to market incentives, the economically efficient ideal would be to manage the forest for its multiple uses by intertemporally maximising the net social value from all the jointly produced outputs of the forest (Krutilla and Bowes 1985). Such management would properly account for all the externalities in a context of an underlying multiproduct production function, with its complementary and competitive features.

IV. GOVERNMENTAL INTERVENTIONS: AMERICAN EXPERIENCE

Background

The Europeans colonising America found vast natural forests. For all intents and purposes, these forests were predominantly an open access resource. While much was ostensibly owned by the state, the outputs of the forest were generally available for use to any who would bear the cost of gathering or harvesting. While the forest was an asset to the colonists, it also was an impediment to other land uses and much of the early history of American forests involved removal of the forests to make way for other uses -- primarily agriculture. Clearing proceeded slowly for many decades.

As late as 1800 only a few million hectares of the original forest had been cleared. The process accelerated in the 19th century, especially the latter part, probably reaching its peak around 1905 (Clawson 1978).

Toward the latter part of the 19th century a growing concern began to emerge over the destruction of the natural forest by "timber barons" and others and its consequences for future timber availability. Because of this concern, national "forest reserves" were established in the late 1890s (the beginnings of the National Forests). The Forest Service was created in the early 1900s to provide stewardship and management of the National Forests.

The passing of management and control of large areas of forest to permanent public ownership reflected a growing disillusionment with the performance of the private sector in its management of the forests. The nationalisation of large areas of forests in the United States reflected concern that "market failures" were pervasive in forestry. A benign public agency was created to optimally intervene and manage in a fashion so as to correct the distortions created by the market.

V. SOME PROBLEMS IN PUBLIC SECTOR FORESTRY

While the concept of government intervention of various forms in the forestry sector to improve its economic performance seems reasonable in light of the sector's potential for external effects, the actual performance of government as a regulator and especially as a manager of public forests has been subject to a great deal of criticism from analysis. Some of those criticisms in the American case are listed below.

In the United States, public involvement in forestry occurs in at least two forms. First, there is direct government ownership of forestlands. The National Forests, discussed above, involve the largest area, but there are other public owners of forests at the federal, state, and local levels. Second, there is public regulation of private forestry activities such as might be embodied in "forest practices" acts.

Below-Cost Sales

The experience of federally-managed forests in the United States has been mixed. While for many years the Forest Service was considered to be one of the elite public agencies, in recent years that reputation has been somewhat tarnished as the management policies of the Forest Service have come under criticism from a growing number of sources.

A major contentious issue in public forestry in the United States is what is known as "below-cost timber sales" or deficit sales. This situation occurs when the costs of harvesting the timber on public lands exceed the revenues from the timber sales. In the context of a private decision, such a harvest would never occur unless some other associated benefits were identified and included in the timber harvesting calculation. Below-cost sales have been common in many of the National Forests for years. The rationale is usually either a) that there exist "timber dependent" communities and this indirect subsidy is therefore justified, or b) that there are some other timber management benefits which, when considered with the timber value, justify the harvest. An example of such a benefit would be wild animal browsing.

This practice has come increasingly under attack in recent years by environmentalists who argue that the effect of these sales is to subsidise the destruction of natural forest. The argument that beneficial associated effects are generated is also frequently challenged by these groups which maintain that the associated effects are commonly minimal and often negative. One notable example of this policy is the Tongass National Forest in Alaska where environmentalists claim timber is being harvested for a return of five cents on the dollar and in the process a scenic wonder is being destroyed. (Wilderness Society 1986).

While it is likely that in some cases non-timber benefits would justify below-cost harvests, it also seems obvious that a large number of these sales can never be justified on economic or social efficiency grounds. This appears to be a clear example of governmental failure and an area in which market performance motivated by market incentives would likely generate a socially superior outcome. The reasons for the continuation of harvests in many of these regions appear to be explained best in terms of Public Choice theory and the budget maximisation hypothesis (Johnson 1985).

State Forest Practices Acts

The argument of externalities and market failure has led to a variety of regulations in the United States. In a recent study Boyd and Hyde (forthcoming) examine a number of these regulations to empirically determine their effect. One of their investigations compares the performance of the private forestry sector in two adjoining states -- Virginia and North Carolina -- one with (North Carolina) and one without state regulations on reforestation. Their statistical investigation finds no evidence that the North Carolina law has been effective.

More generally, after an examination of several regulations in the forestry sector, Boyd and Hyde conclude that the economic justification for each individual regulation is elusive, and that what was originally justified on grounds of market failure subsequently winds up being justified by its distribution impacts. This view is quite consistent with the notion of various rent-seeking agents as developed originally by Krueger.

Public Investment

Associated with the management of the National Forests is the need to allocate large amounts of public funds to address investment needs. In the United States the allocation of investment funds has been under heavy criticism. A major criticism has been the lack of economic criteria (broadly defined to include external effects) in the investment allocation decisions (Clawson 1976). For example, it is alleged that the Forest Service has allocated funds for reforestation in inverse relationship to the economic returns to the activity. Variations of the criticism of inadequate economic criteria are found with respect to rotation lengths and the broader questions of allowable cut and appropriate drawdown of old growth timber (Dowdle and Hanke 1985, Lenard 1981).

Outside the United States criticisms of the economic efficiency of government management of forests is also heard. Two examples are given below.

Canadian Stumpage Fees

In Canada, most of the forestland, particularly in the west, is owned by the provinces. The usual procedure for allocating harvesting rights is by a long-term contract worked out between the provincial government and a forest products company. A common criticism of these arrangements is the lack of competition in the bidding for rights and the administrative formula method, rather than a market method, of establishing the charge for the stumpage rights.

Harvest at Culmination of Mean Annual Increment: A European Practice

A common criticism made by North American economists of much of public forestry management in Europe is the practice, common in government and government regulated forests, of using the foresters' biological rule for determining the length of the harvest rotation. As economists point out, the mean annual increment rule was demonstrated to be economically inappropriate in 1849 by the German forest economist Faustmann, yet the practice seems to continue.

VI. WHY GOVERNMENTAL FAILURES?

In the face of an apparently compelling theoretical case for government intervention to offset "market failures" in the forestry sector, the "failures" of government management and regulation listed above may come as a bit of a surprise. Why might governments "fail" in their management and regulation of forestlands? More generally, are these failures unusual, are they inherent in the present system or are they more broadly inherent in government?

There are at least two types of reasons given to explain government failures (Krutilla et al. 1983). The first is the argument that the public sector does not adequately take economic factors into account in decision-making. The approach suggested to remedy this is the introduction of better techniques and planning models. The second is the view that the public sector lacks appropriate incentives to bring about an efficient allocation. Market mechanisms and information are missing from the public decision-making process. To remedy this second source of government failure, a host of changes, from reform to radical restructuring, have been proposed. Many of these changes would take the management decision away from the public sector altogether. All aim at increasing the degree to which market incentives and information are incorporated into the forest management decision-making.

A third broader critique made by the Public Choice School maintains that public sector activities invariably result in "government failures" due to the distorted set of incentives facing the administrator and bureaucrat. Hence, Public Choice theorists would be very skeptical that a government agency could effectively carry out its responsibilities, even if it had adequate physical and economic data and if economics was used more broadly.

Lack of Economic Analysis

The first criticism, that of not adequately utilising economics and markets, is a common criticism of public resource management agencies and not confined to forestry. In forestry, this view is simply that more economics needs to be introduced into the government's management, planning and decision-making. Improvements in the extent of economic analysis have surely taken place. Today, in the United States the large federally owned forests are required by law to be managed for multiple-use purposes. This mandate has been viewed as approximating the maximisation of economic value, broadly defined (Krutilla and Haigh 1978). A number of economic models have been developed for utilisation in public forestry management and the Forest Service has drawn up regulations that increase the role of economic considerations in public forestry management.

However, while it is certainly true that there are a considerable number of situations where economics can usefully be introduced into public agency management of forests, there are still reasons to believe that this may not be adequate for ensuring consistently efficient management. The lack of appropriate pricing for environmental values creates uncertainly as to the magnitudes for the various outputs. In many cases the "prices" introduced into the economic calculus are highly debatable. While "benefits studies" are sometimes used to estimate the magnitudes of the values of non-marketed outputs, they lack the definitive feature of a market price and are often disputed by contending groups. For example, one criterion for the rate at which to harvest old growth forest in the Pacific Northwest is in large measure the "externality" value of the old growth as an amenity. Environmental groups argue, in essence, that this externality is so great that the old growth should never be harvested. In addition, the latitude for pure economic decision-making is often highly constrained, even in planning models. Thus, an economic model may be so highly circumscribed by non-economic constraints that the "solutions" say little about economic efficiency (Walker 1983).

Furthermore, while work has proceeded on multiple use models, the results are sometimes discouraging in terms of providing useful guidance for public forest managers. Perhaps more importantly, the lack of scientific knowledge as to the actual extent of the underlying inter-relationships between the several forest outputs makes the composition of the economically efficient optimal output mix highly conjectural. While it is sometimes argued that large non-timber benefits, e.g. animal browsing and recreational access, are created by a particular harvest, the pervasiveness of these values has been challenged (LeMaster et al. 1987). In an extensive study of the economics of multiple-use forestry in the context of important inter-relationships between timber stands, harvests and other benefits, Bowers and Krutilla (1985) state:

"The general multiple-use harvesting policy is seen to be complex. No simple rule of thumb is likely to describe the harvest. We see that sometimes younger stands are harvested leaving older ones uncut. We may choose to briefly delay regeneration. We rarely cut a particular stand at the same age twice in succession during the initial periods. The forest areas may be managed with some areas set aside for specialised purposes -- old growth or clearing for wildlife."

They conclude:

"Perhaps, most importantly, we see from these examples, that the harvesting decision can be extremely sensitive to factors about which we have little empirical knowledge (p. 566-67 emphasis added)."

The dilemma that arises is as follows: if externalities are large and pervasive in forests then governmental intervention is justified; however, if the relationships are as complex as suggested above and if the empirical knowledge is lacking, it is difficult to see how the government can intervene effectively to optimise management, even in the best of circumstances. Finally, even if research reveals these complex empirical inter-relationships for a particular site, the costs of developing sufficient information to plan and manage large public forests in accordance with socially maximising principles are likely to be prohibitive. Achieving "ideal" management in this way may be inefficient in that the information costs exceed the benefits from improved management. A less "ideal", but less costly management approach may prove to be socially more efficient.

Restructuring the Public Sector to Incorporate Market Incentives

This criticism calls for broad institutional changes and the purpose of these institutional changes is ultimately to introduce more market forces and economic incentives into the management decision-making process. These changes run the gamut from minor reforms to the outright privatisation of the bulk of public forests (Gardner 1983). One variant calls for the utilisation of "the adaptive properties of the markets themselves as a planning mechanism" by placing individual public forests into "profit centers" under the Forest Service (Binkley 1983). Between the extremes are various proposals for extending long-term leases to various groups with different management goals. Whereas the more modest of these proposals would require civil servants to respond to market incentives, the more radical involve direct profit-making incentives so that appropriate decisions are rewarded by the market in profits and capital appreciation and incorrect decisions are penalised by economic losses.

These proposals to make managers more responsive to market forces and economic incentives are akin to the view presented above in the paper on water by Frederick. In water, for example, it is argued that greater use of market exchanges would enhance economic efficiency. In forestry, the arguments take the form of valuing extra market resources by a set of administratively set "transfer prices". More generally, these views call for forest management by groups having a strong interest in the outcome of their management decisions and hence strong incentives to make "correct" decisions.

While the incentives typically relate to monetary values, they need not do so exclusively. For example, some advocate that long-term control of wilderness areas be transferred to wilderness groups. The goal of this restructuring of control is to place the costs and responsibility of the output promoted directly on highly interested parties. For example, if wilderness areas also provided mineral potential, manager/owners would have real monetary economic incentives to develop the minerals while continuing to safeguard their personal values in protecting the wildlife and/or other wilderness amenities. Thus, opportunity costs and trade-offs need to be addressed directly by the manager. Revenues from minerals could be used to support the cause of wilderness groups. Control by a wilderness groups rather than government would make the decision-making process less political with the decision-makers facing more directly the actual social costs and trade-offs involved in management and decision-making, especially where more than one important value is involved.

The Public Choice Critique

In recent years the argument of "government failure", i.e. failure of the government to manage or regulate in a socially desirable manner, is a frequent criticism by the Public Choice School. The concept offers an effective counterweight to the concept of "market failures". Why might government fail to intervene optimally to counteract the failures of the market and achieve socially desirable results? The reasons given by the Public Choice School arise principally out of political allocations and various types of inefficient non-price competition (Gardner 1986). While it is often argued that the government will take a long-term time perspective, this view argues that the political and administrative processes are often quite myopic and hence may fail to properly consider the long-term implications (Stroup 1983).

A related point raised by the Public Choice theorists (Buchanan and Tullock 1962) is that governmental decisions are often driven by considerations other than social welfare. Thus Niskanen (1971) argues that agencies and bureaucrats have parochial interests that, at least in part, are different from the mandated objectives of their agencies. These considerations suggest that public and administrative decision-making is likely to be systematically flawed and hence government failure the rule rather than the exception.

Finally, an additional criticism is raised. In a rent-seeking environment, i.e. an environment where large transfers of wealth can be obtained by influencing the political and administrative decision process with regard to publicly managed resources or public regulations, the potential rents from the resource will often be highly dissipated in the process of trying to acquire some of the returns to that asset (Deacon and Johnson 1985, Krueger 1974). This final point argues that just as an open access common property resource will tend to be overexploited and wasted, so too, resources will also be wasted in the process of trying to influence decisions so as to allow special interest groups to capture rents from a publicly-managed resource.

Does the Public Choice criticism apply to forestry? What is the empirical evidence in the case against public sector ownership and management in the forestry sector? While all of the evidence is by no means in, a number of recent books published in the United States document a strong empirical case arguing that the publicly owned and managed forestlands are being badly managed (Deacon and Johnson 1985, Emerson, forthcoming), with numerous case studies provided. Furthermore, public policies designed to impact on the forestry sector more generally are having little or none of the desired impacts (Boyd and Hyde, forthcoming).

While much of this criticism does not come from a Public Choice perspective, it does tend to be consistent with that perspective. These results also complement the earlier criticisms of Clawson and others and provide a rationale as to why economics are not used more efficiently in the management of public forests. These results are all the more startling when it is realised that the United States Forest Service is recognised as one of the most technically competent and dedicated agencies in the United States government. If United States public forests are badly managed, the Public Choice argument goes, it is due not to technical incompetence or lack of well-intentioned civil servants but rather to the inherent difficulties of public sector management and intervention.

VII. PRIVATE INCENTIVES FOR TIMBER PRODUCTION

Timber Production

There seems to be relatively little doubt, at least in the United States, that the private sector and the market can do an effective and efficient job of producing industrial wood. Experience suggests that private sector markets tend to be efficient in producing commodity outputs and timber is no exception. The private market calculation for timber-growing, like other investments, compares costs and returns adjusted for time and the discount rate. In an earlier study (1983), the present author found that the economic returns to timber growing were quite favourable in a number of locations worldwide. In general, the wood-producing plantation forests of the future are likely to be found on high productivity sites with gentle terrain and good access. These conditions not only hold production costs down but also tend to be areas where environmental problems are likely to be modest and forest amenity values small.

When dealing with an existing stock of mature timber or old growth timber, the economics again are quite clear for the private timber producer.

The tree should be harvested when the opportunity costs of not doing so exceed the gain of a delay in harvest. Considerations here include current and anticipated real timber prices, the discount rate, the biological growth of the trees and the lost opportunity of putting the site into new rapidly-growing seedlings.

As with other commodities, market signals provide firms with information in the form of prices and costs. These translate into incentives in the form of profits if the appropriate decisions are made. Of course, as with other economic activities, uncertainties exist, particularly with regard to the course of future prices and costs. However, these uncertainties exist for public decision-makers as well as private and there is little reason to believe the public sector can forecast future prices any better than does the private sector. In any event, forecasts made by public agencies can be made available to private sector decision-makers.

In the United States, the private sector produces the majority of the nation's industrial wood. Furthermore, with the diminution of the old growth timber reserves, the future timber supply will come increasingly from second growth and plantation forests. The private sector has been much more active than the public sector in establishing forest plantations and management for timber values. For example, while in 1985 the private sector planted 900 000 hectares in forests, the public sector planted only 100 000 hectares. While public sector financed incentive programmes for private lands have no doubt had some impact, Boyd and Hyde (forthcoming) argue that such impacts have been small and often inefficiently achieved.

Non-timber Outputs of Forests

If the forest sector is to experience major market failure, it is likely to fail not in its production of industrial wood, but rather as a result of inadequate attention to the nonwood outputs of the forest including environmental services. Besides wood, forests provide outputs in the form of amenity values such as recreation and wildlife, habitat and in the form of environmental services such as watershed protection and erosion control. It is in the provision of these services that the market is more likely to experience failure. While recreation and wildlife, particularly in the form of hunting services, are marketable commodities, often society wants to treat these as "merit goods", goods that society should receive essentially at no cost to the individual. The market is unlikely to provide this if it involves extra costs to the producer.

In many cases, however, markets for these services have in fact emerged and forest owners have often been quick to respond to the incentives of these markets to provide for fee-paying recreation and hunting. In the United States, at least, private camping grounds are found in all parts of the country, often closely in competition with public facilities. Also, fee-paying hunting on private lands has grown substantially in recent years. A recent study by Lassiter (1983) documents the rather substantial expenditures that private woodland owners are incurring in the process of improving the wildlife on their woodland in southern United States. A major incentive for this activity is in the revenues generated via the selling of hunting permits and the leasing of hunting rights to groups and hunting clubs.

In the case of environmental values, however, private markets typically do not emerge. However, conflicts between industrial forestry and these values, e.g. watershed protection and erosion control, are probably rather small in a large number of cases (LeMaster et al. 1987, Sedjo 1986). This is likely to be especially true for the high-productivity plantation forests that are found in relatively flat accessible areas such as southern United States. For example, plantation forests as well as natural forests provide watershed protection, erosion control, and other environmental services as a by-product of their operation.

Since tree-growing involves mostly growing and only occasionally harvesting, conflicts between commercial forestry harvests and environmental protection are likely to occur only infrequently. Thus, it may well be the case that as a practical matter the inter-relatedness and negative external effects are more modest than sometimes believed.

It has been stated that the question of whether market failure or government failure is more serious is ultimately an empirical question that depends upon the particular situation (Gardner 1985). If it is correct that most gentle terrain commercial forestry operations and non-timber values are not highly in conflict, then harvests approximating private schedules might also approximate the social ideal for many forests. Furthermore, even at harvest, the extra costs of making a harvest environmentally safe are likely to be modest in most cases. Thus, while the theoretical "ideal" of optimising the outputs in a multiple use context may not be achieved, the conflicts in most forests are likely to be modest and manageable with reasonable performance based upon response to market incentives and perhaps some form of modest "foresty practices" regulations.

VIII. A ROLE FOR THE PUBLIC SECTOR IN FORESTRY

The potential conflicts between harvesting and environmental values are likely to be greatest on steep slopes and in difficult terrain. Fortunately, this type of terrain typically does not lend itself particularly well to commercial forestry. The difficult terrain makes harvesting, planting, and other activities costly, and hence discourages much large-scale industrial forest activity. Thus, as noted earlier in the below-costs discussion, in many cases the private market is not willing to incur the costs associated with access and harvest costs from difficult terrain.

However, while there is generally little chance that market incentives would dictate this type of terrain as a major timber-producing area for the reasons mentioned above, incentives may exist for a "cut-and-run" timber mining activity. That is, the market incentive may justify extraction but not reforestation. Historically, in the United States many of these areas have become part of the public lands. Where scenic or natural settings are unique, these areas have often been set aside as National Parks, wilderness areas, and the like. The rationale for public ownership is the perception of unique natural values that are seen as not readily marketable in an unadulterated form. It is here that the traditional argument of market failure is made and the use of public intervention is justified to offset that failure.

In the presence of a market failure, the options appear to be either to rely on governmental interventions that are likely to result in government failure, or to be prepared to accept the original market failure. Public policy presumably should try to provide for the mix of public and private decisions that minimise the net cost of the distortions due to "failures" of both types.

Consistent with this aim, Nelson argued (1982) that in the United States the private sector ought to be almost exclusively responsible for commodity production (industrial wood), while the public sector should be given responsibility for maintaining many recreational lands, particularly unique scenic wonders, mountaintop parks, and the like. Nelson extends this concept to propose that United States public lands be designated as either for timber production or for recreational use. The land designated for timber production would then be managed by the private sector while the recreation lands would be managed by the government for recreational and environmental outputs.

IX. SUMMARY AND CONCLUSIONS

Forests produce a variety of outputs. Wood as a commodity poses no special problems for establishing efficient markets. Property rights are well-defined and easily transferable, and markets for timberlands, stumpage and logs are common. However, while forest ownership confers property rights to land and trees, it typically does not confer ownership of environmental outputs of the forest such as water and watershed protection. Environmental outputs typically lack well-defined property rights and markets.

Ideally, the optimum utilisation of the forest would involve consideration of the jointness in production of the various outputs, commodity and non-commodity, and maximise the net social value of all the outputs. Private markets cannot accomplish this, at least not in the absence of markets for some outputs and the existing distribution of property rights. This potential failure of the market provides the rationale for public forestland ownership.

Governmental interventions to offset market failures, however, have been plagued with inefficiencies and corresponding "government failures". While "market failures" are due to the lack of clearly-defined property rights and the resulting externalities, "government failures" in forestry have been attributed to several causes. First is the lack of proper and sufficiently detailed economic analysis and planning models. Second is the failure to utilise adequately market incentives in the public management and regulatory process. Third is the broad criticism of the Public Choice School which argues that the incentives in public management are inherently flawed and distorted.

This paper argues that, for much forestland, the conflicts between industrial forestry and environmental and other outputs are small and most forest sites do not seem to have the serious externalities associated with "normal" forestry practices. Therefore, it is probably more efficient to simply allow private ownership to manage the forest for individually-determined objectives and address serious externalities by selective adaptations to accommodate them where necessary. Private forest management, or "private-like" management in the context of regulations to mitigate serious environmental damages, is likely to provide acceptable performance in the production of both commodity and environmental outputs.

Finally, in the natural system it is often true that where environmental and recreational values are large, timber values are small. In this context, public ownership with management for environmental and recreational values may be most efficient.

NOTES AND REFERENCES

1.   An interesting example is found in the forests of Nepal. Until about
      1955 the forests were the property of the villages and the villages had
      developed communal systems for their protection, exploitation and
      regeneration. However, in 1955 the central government made the forests
      the property of the central government and with it the responsibility
      of protection. The shift of ownership to the government was viewed by
      the villages as expropriation. The villages no longer viewed the
      forest as their property and in so doing lost incentive for its proper
      long-term management and protection. The government lacked the ability
      to protect and manage the forests effectively. Poaching from the
      forest became common place. Given the government's inability to
      provide even minimal protection, the forest became an open access
      common property resource. Villages excessively exploited the resource
      since they recognised that if they did not, someone else would.
      Incentives for long-term management and investment in the forests'
      future disappeared. Some analysis attributes the rapid deterioration
      of the Nepalian forest after 1955 to the simple act of government
      expropriation. By the 1980s some attempts were being made to reverse
      this process with the reinstating of village owned and managed
      woodlots. However, by that time a great deal of damage had already
      been done.

2.   Binkley, Clark S. 1983. "Comments" in Governmental Interventions,
      Social Needs and the Management of U.S. Forests, edited by R.A. Sedjo,
      Published by Johns Hopkins Press, Baltimore, for Resources for the
      Future, pp. 237-44.

3.   Bowes, Michael D. and John V. Krutilla. 1985. "Multiple Use Management
      of Public Forestlands", in Handbook of Natural Resource and Energy
      Economics, vol. II, edited by A.V. Kneese, J.L. Sweeney, Elsevier
      Science Publishers, Amsterdam.

4.   Bowes, Michael D., John V. Krutilla and Paul B. Sherman. 1984. "Forest
      Management for Increased Timber and Water Yields", Water Resources
      Research, June.

5.   Boyd, Roy G. and William F. Hyde. 1987. Public Regulation of a Private
      Forest Resource, forthcoming, University of Iowa Press, Iowa City.

6.   Buchanan, James and Gordon Tullock. 1962. The Calculus of Consent,
      University of Michigan Press, Ann Arbor.

7.   Chapagain, Devendra P. 1985. "Nepal: A case study of Alternatives in Resource Management", paper presented to the seminar on "Managing Renewable Resources: History and Contemporary Perspectives", sponsored by the Japan Center for International Exchange and the Agricultural Development Council, Sapporo, Japan, June 24-25.

8.   Clawson, Marion. 1976. "The National Forests", Science, Vol. 191, No. 4228, pp. 62-67.

9.   Clawson, Marion. 1978. "Forests in the Long Sweep of American History", Science, Vol. 204, No. 4398, pp. 168-74.

10. Clawson, Marion. 1985. "Problems of Public Investment in Forestry", in Investments in Forestry, R.A. Sedjo editor, Westview Press, Boulder, pp. 167-85.

11. Coase, Ronald. 1960. "The Problem of Social Cost", Journal of Law and Economics, 3 (October), pp. 1-41.

12. Frederick, Kenneth. 1987. "Water Resource Management and the Environment" (this volume).

13. Gardner, B. Delworth. 1984. "The Case for Divestiture", in "Rethinking the Federal Lands", editor S. Brubaker, Johns Hopkins Press, Baltimore, for Resources for the Future, pp. 156-180.

14. Gardner, B. Delworth. 1985. "Foreword", to Forestlands: Public and Private, editors R.T. Deacon and M.B. Johnson, Ballenger Press for the Pacific Institute, Cambridge, Mass.

15. Hardin, Garett. 1968. "The Tragedy of the Commons", Science, 162, 1243-8.

16. Krueger, Anne O. 1974. "The Political Economy of the Rent-Seeking Society", American Economic Review, June. pp. 291-303.

17. Krutilla, John V. and John A. Haigh. 1978. "An Integrated Approach to National Forest Management", Environmental Law, Vol. 8, No. 2.

18. Krutilla, John V., Anthony Fisher, William F. Hyde and V. Kerry Smith. 1983. "Public versus Private Ownership: The Federal Lands Case", Journal of Policy Analysis and Management, Vol. 2, No. 4, pp. 548-558.

19. Johnson, Ronald N. 1985. "U.S. Forest Service Policy and its Budget", in Forestlands Public and Private, editors R.T. Deacon and M.B. Johnson, Ballenger Press for the Pacific Institute, Cambridge, Mass.

20. Lassister, Roy L. Jr. 1984. "Access to and Management of the Wildlife Resources on Large Private Timberland Holdings in the Southeast United States", College of Business Administration Monograph Series - No. 1, Tennessee Technological University, Cookeville, TN.

21. LeMaster, Dennis C., Barry R. Flamm and John C. Hendee, editors. 1987. Below-Cost Sales: A Conference on the Economics of National Forest Timber Sales, Wilderness Society, Washington, D.C.

22. Lenard, Thomas M. 1981. "Wasting Our National Forests", Regulation, July/August, pp. 29-36.

23. Nelson, Robert H. 1982. "The Public Lands", in Current Issues in Natural Resource Policy", editor P.R. Portney, Johns Hopkins Press, Baltimore, for Resources for the Future, pp. 14-73.

24. Menzies, Nick. 1985. "Land Tenure and Resources Utilization in China: a Historical Perspective", presented to the seminar on "Managing Renewable Resources: Historical and Contemporary Perspective".

25. Osako, Masako M. 1983. "Forest Preservation in Tokugawa Japan", in Global Deforestation and the Nineteenth-Century World Economy, edited by R.P. Tucker and J.F. Richards, Duke Press Policy Studies, Durham.

26. Sedjo, Roger A. 1983. The Comparative Economics of Plantation Forestry: A Global Assessment, Johns Hopkins Press for Resources for the Future, Baltimore.

27. Sedjo, Roger A. 1987. "Below-Cost Timber Sales: A Summary", in Below-Cost Sales: A Conference on the Economics of National Forest Timber Sales, editors D.C. LeMaster, B.R. Flamm and J.C. Hendee. Wilderness Society, Washington, D.C.

28. Sedjo, Roger A. and Marion Clawson. 1983. "Global Forests", in The Resourceful Earth, editors Julian Simon and Herman Kahn, Basil Blackwell Inc., New York.

29. Stroup, Richard. 1983. "Comments" in Governmental Interventions, Social Needs and the Management of U.S. Forests, edited by R.A. Sedjo, Johns Hopkins University Press, Baltimore, for Resources for the Future, pp. 115-123.

30. Walker, John L. 1983. "National Forest Planning: An Economic Critique", Governmental Interventions, Social Needs and the Management of U.S. Forests, edited by R.A. Sedjo, Johns Hopkins Press for Resources for the Future, pp. 263-296.

31. Wilderness Society. 1986. "America's Vanishing Rain Forest", April.

Chapter 6

THE ROLE OF ENVIRONMENT IN FOREST RESOURCE MANAGEMENT

Dr. John Hosteland

I. INTRODUCTION

Although Sedjo (Chapter 5) examined the history of forest utilisation and management largely in the United States context, much of his reasoning has global relevance. Even if this paper concentrates on experiences in forest management in Scandinavia, and particularly in Norway, it might be helpful to start by quoting an American book called "Up for Grabs. Inquiries into who wants what" by Daniel Jack Chasan. The first chapter of his book is called "The Last Tree" and starts as follows:

"Not long ago, a friend of mine talked with an old man who claimed to have cut down the very last big virgin Douglas fir on Vashon Island, where I live. The tree was some two hundred feet tall, the old man said, and it rose one hundred feet before the straight sweep of the trunk was broken by the first branches. It was a good thirteen feet across at the base. Much of the island is still covered with trees, but not trees like that. My friend had an acquaintance in the forest products industry do a little quick calculating, and found that there had probably been enough good lumber in that tree to frame twelve houses. But the old man hadn't cut it for lumber to build houses. He had cut the last virgin Douglas fir on Vashon Island for firewood. It made good firewood, too. He burned it in his iron stove for ten years.

It seems absurd, from a later perspective, to have cut that tree for firewood. The man probably had little money at the time -- it was the Depression -- and could certainly be forgiven the desire to keep himself and his family warm. But it was a disaster. If you think of the tree as a tree, it was an aesthetic disaster. If you think of the tree as lumber, it was an economic disaster."

This story tells us quite a lot about the problems we face when utilising natural resources. Whether we do so efficiently will be viewed differently at different times and by different interest groups.

Theodore Roosevelt once defined conservation as "wise use" of resources. Everyone thinks he or she wants to use a given resource wisely. But what looks like wisdom to Greenpeace may look like folly to the forest owner or the forest products industry, and vice versa. Clearly, there are people who resent the loss of every tree in any forest. Clearly, they are not the people who earn a living by logging, working in lumber mills or building frame houses. Nor are they the executives of forest properties or forest products companies. Whose definition of efficiency applies? Who gets to decide who benefits?

Each society has worked out its own answers to these questions. Each has its own range of answers derived from its specific culture and economy. But the questions apply in every society.

One need not go back far in history to find a time when there was no concern whatsoever about negative environmental effects of forestry practice. The forest has been utilised for timber-harvesting for hundreds of years in Scandinavia. Periodically, there has been concern about over-utilisation and depletion of the resource, and political measures have been taken.

Regulations on cutting of young stands and the obligation to reforest after harvest were introduced at an early stage. In Sweden for example there was heavy harvesting of oak for shipbuilding from 1500 onwards. In fear of short suppy, a law was introduced around 1600 which said that for every large oak tree felled, at least 10 new trees should be planted. The oak-trees planted at that time are mature for cutting today.

With the long tradition of forest utilisation in Europe, the norms for what is "good forestry practice" or "wise use" of the resource have developed. Forest policy generally focussed on maximising wood production. But equally important has been legislation and economic incentives to conserve, establish and improve forest resources. Forestry practice is however somewhat different in Europe compared to the United States.

II. NEW TECHNOLOGY, CHANGING NEEDS

There have been two important developments which make our concern for the environment more pressing today than earlier. First, new technology has changed forest operations and produced environmental effects increasingly perceived as negative. Secondly, economic development has led to a new and increasing demand for "environmental services".

These two factors have created conflicting views on the efficiency of the utilisation of forest resources. The relative values of different services and commodities produced by forestry has changed. The problem is that while timber is efficiently priced in a market, price signals for environmental services or outputs are lacking.

Moreover, benefits as perceived by one group may represent costs for another group. Where decisions are based exclusively on an evaluation of costs and benefits, it is obvious that the decision process may be inefficient.

When examining this situation, we often concentrate on external costs that are not internalised in the management of timber production, and it is easy to agree that it is important to recognise environmental factors. But the consumption of recreational services for example raises problems in that the lack of a price (cost) can lead to inefficient decisions. No price, or a price equal to zero, will theoretically lead to over-utilisation.

III. RELATIVE IMPORTANCE OF FORESTRY OUTPUTS

Sedjo concludes in Chapter 5 that the conflicts between industrial forestry and environmental and other outputs are small. Small is a relative word so it easy to agree in most respects. But it is important to recognise the conflicts that do exist and to be aware that they may worsen if we do not cope with the situation in an efficient way.

The importance of inefficiencies in the management of forest resources differ from country to country and between regions within each country. Even if the approach to management of forests is the same, the conflicts will differ due to different demands for environmental benefits. This supports the conclusion by Sedjo that serious externalities should be addressed by selective policy measures.

Sedjo also states that when environmental and recreational values are large the timber values are small. That may generally be the case, but as environmental and recreational values very often are a function of population concentration, it will not always be so.

The visual landscape quality of the forest is often very high in these areas, and on an attractive site, the effect of modern forestry practices will often produce more dramatic changes in the landscape than in less attractive areas.

IV. FOREST POLICY IN NORWAY

The conflicts that may and do exist between modern forestry and other use of the forests are currently addressed in different ways. In Norway both legislation and economic incentives have been developed to promote forest resource utilisation that maintains a balance among the different objectives.

The political authorities realise that the forest plays a more and more important role as a resource for recreation, the natural environment, and science. At the same time it is still a primary aim that forests should be exploited efficiently in order to produce sufficient industrial wood, and to maintain satisfactory living conditions for rural people. It should be noted that in Norway more than 80 per cent of the forest area is privately owned. In Sweden and Finland, most forest land is also in private hands.

The Norwegian authorities try to balance conflicting interests by amendments to the Forestry Act. The objective of the Forestry Act is to promote forest production, afforestation and forest protection. It aims to obtain satisfactory results for those employed in forestry. Furthermore, stress is laid on the importance of forests for recreation, as an integral part of the landscape, as an essential environment for plants and animals and as an area for hunting and fishing.

This focus on multiple use was reflected in the Forestry Act already in 1965, and forest management for multiple use has since then become more and more important.

The underlying principle of the Forestry Act is that the forest owner should be free to manage his forest without intervention from the authorities as long as this is done in accordance with the principles of sound forestry practice.

Thus, the Act contains guidelines for marking, felling, planting, regeneration and silviculture which take into account forest protection, recreational activities, hunting and fishing. Furthermore, the Act gives the Forest Service authority to intervene in cases of poor management, when forests have been damaged by fire, windthrow, landslide, etc., or when forests are threatened by disease or injurious insects.

The Ministry of Agriculture may decide that the forest shall be classified as a protection forest against landslide, avalanches or other natural disasters. The rules for silviculture in such forests are more severe than for an ordinary forest.

In recent years there has been considerable debate as to whether a conservation or multiple-use approach is the best way to solve the conflicts between timber production and environmental outputs. Interest groups claim that the only way to guarantee the environmental and recreational values of an area is conservation. The argument is that as long as traditional forest management is profit motivated, it is impossible to arrive at an optimal mix for both non-marked goods like recreational values and wood production.

Forest owners on the other hand maintain that timber production is so important for the country that it must remain the main objective of forestry, and that "good forestry" is also good environmental management. One of the big issues regarding conservation is compensation to the forest owner. Rules for compensation depend on the extent of the restrictions on forestry practice, but it is always difficult to evaluate the economic consequences for the forest owner. Another issue has been the question of what restrictions must be accepted by the forest owner without compensation.

A further problem is the widespread attitude that the forest should be managed like "in the good old days", meaning very extensive forestry techniques with minimal use of modern equipment. This may be impossible for a forest owner who has to live from the income he can extract from the forest.

When an area is conserved for wilderness, it develops rather rapidly into a primeval forest which is not very suitable for recreational use. The best way in most cases is to try to develop the multiple use approach and to develop economic incentives to achieve this goal. A general law will not suffice, when economic realities are the main basis for forest management.

V. ECONOMIC INCENTIVES

There are three types of economic incentives in Norway:

-- Public grants
-- Tax incentives
-- Forest Levy (compulsory investment fund).

Most of these have been established to promote forest production, but today several conditions involving environmental considerations have been attached to different incentives.

Public Grants

To enable forest owners to implement agreed forest policy different types of public grants are proposed. The application for and the use of grants, as well as disbursements, are controlled by the district foresters of the Forest Authority.

When subsidies are used to promote a specific objective, the question arises of whether such interventions by the authorities can in themselves cause inefficient management in terms of environmental effects.

One finds a good example of this when looking at the effect of subsidisation of afforestation in the western parts of Norway. The objective has been to increase production of spruce in an area where only birch grows naturally. Because a spruce forest is so different from a birch forest the effect on the landscape has in many cases been dramatic. In this mountainous region with small properties, the result has been dark green dots scattered over the mountain side. We talk of "square forests" -- and that is a negative term.

Today one is much more concerned about these effects. When an area is afforested, the total effect on the landscape should be taken into consideration. This is possible when the Forest Authority at the local level provides both the right directives and controls the financial resources. (Between 70 and 80 per cent of the cost of afforestation is paid by the Government as grants.)

Road construction is another area where we experience certain conflicts between timber production and recreational activities. In most cases the Government subsidises road construction (40 - 60 per cent of the cost). Very few roads are built without subsidies and therefore the Government has a great deal of influence on the construction.

In the most sensitive areas all new roads must be approved by the Forest Authority and the forest owner must apply for a concession to build roads.

With such a system of both economic and legislative measures, society can combine to a great extent the interests of both environment and forest production.

Management plans for forestry are also subsidised by the Government to various degrees. This is done both to encourage forest owners to have a management plan, but also to have the opportunity to influence the content of the plan. Earlier the plans' main focus was on cubic meters, age distribution and site quality. Today, an attempt is being made to integrate management strategies for wildlife habitats and recreational values. Even if the economic consequences of managing the forest for recreational values can be negative for the forest owner, the cost will often be very small when it is planned well in advance. The positive effect is less conflict with the public.

As stated earlier, it is important to develop norms for what is "good forestry practice", and this is one way to do that.

In Sweden it is compulsory to have a management plan when the forest property is over a certain size. These plans are important for developing the wisest possible use of forest resources. Management plans may provide scope for an even better integration of environmental concerns and other management objectives.

Tax Incentives

The taxation system for forestry varies widely among different countries especially as regards succession duties and property sales. It is impossible to go into detail on this subject, but it is obvious that the tax system can have a great influence on forest utlisation and can cause inefficiencies. Generally, tax rules in Norway have not been developed to promote concern for environmental values.

Forest Levy

However, we have the Forest Levy which is a combination of compulsory investment fund and tax benefits. In special areas where environment considerations create extra costs due for example to extensive harvesting methods, these costs can be covered by the Forest Levy. At the same time the forest owner also gets tax benefits.

The Forest Levy was introduced in Norway in 1932. In order to secure sufficient investment in the forest, the Forestry Act provides that the individual forest owner shall set aside a certain percentage of the gross value of all timber sales, i.e. the Forest Levy.

The owner may decide the percentage within a range of 5 to 25 per cent. The Forest Levy is deposited in a special account which gives no interest to the forest owner, to encourage rapid reinvestment after cutting.

The fund may be used to cover expenses for various silvicultural activities, construction of roads and other undertakings in the forest.

Via this fund, the forest owner gains certain tax benefits. A certain percentage of the amounts drawn from the fund is tax free. In practice, the effect is almost like a 35 per cent subsidy. The fund can be used to cover extra costs caused by "environmental management" and this may be of some importance. On paper it seems a good idea; so far it is not apparent that it has had any great effect on forestry practice.

VI. CONCLUSIONS

Efficient resource management is a relative term. What is looked upon as "wise use of resources" changes with time, and when both property rights and markets are lacking, the use of economic incentives alone may not suffice to cope with conflicts of interests.

Sedjo says in his paper that governmental agencies lack proper economic analyses and planning models. This may be true in some respects, but what is probably more serious is the lack of a clear "goal function" and information to put into the models.

If we knew exactly what we wanted and had sufficient information about the consequences of different management strategies, it would be rather easy to calculate an optimal resource use. It is important to remember that when an individual makes a choice between different impulses. When society makes a choice it chooses among competing individuals -- or groups or generations -- and some people inevitably get upset.

The first step towards a better management of forest resources lies in the recognition of different interests and of the negative consequences of certain activities. From there it is possible to search for alternative ways to balance these interests and to cope with inefficiencies. Economic incentives are but one way of tackling these problems.

Chapter 7

HERITAGE MANAGEMENT AND NEGOTIATED MANAGEMENT
IN EUROPEAN FORESTRY

M. Jean de Montgolfier

I. INTRODUCTION

Sedjo (Chapter 5) effectively summarises the current debate in North America over forest management. It is quite clear that the failings he identifies on the other side of the Atlantic are to be found in Europe as well -- selling timber at below its real economic cost, disregard of economic criteria when selecting forestry investment, and government intervention which, perversely puts economic agents on the lookout for grants rather than economic efficiency.

This paper will deal chiefly with three points: a) the different historical and geographical contexts in Europe and North America, a consequence being that in Europe forestry problems cannot be kept separate from overall land use issues; b) the difficulties of taking very long-term costs and benefits into account in economic calculations; and c) the need to transcend the private/public dichotomy and move towards management procedures that are the outcome of bargaining among the actors involved, as part of what we may term heritage management.

II. THE HISTORICAL AND GEOGRAPHICAL CONTEXT OF EUROPE'S FORESTS

As Hosteland (Chapter 6) pointed out, the current situation of Europe's forests cannot be grasped without some outline of their past history.

To begin with, it is now centuries since Europe's forests were common property in the English law sense (the French equivalent would be res nullius -- "gestion en bien commun" -- which has a quite different meaning in French law, as we shall see). In France the last major clear felling dates back to the Early Middle Ages, and since then the whole of France's land area has been held under one form of ownership or another.

In Europe the debate over the roles of private enterprise and central government goes back a long time, though the form and terminology were then different from today's. A few examples can be given, based on developments in northern France. The oldest royal decrees on forestry matters go back to the Middle Ages. Forestry policy in the modern sense dates from the 17th century, with Louis the XIV's Minister Colbert. Timber was becoming scarce, because the best economic return from the forest was given by the production of firewood, grazing and the development of marginal land for crops. Colbert sought to protect timber as a resource, to make France self-sufficient for its shipyards.

With the French Revolution, liberal thinking came to the fore. Many of the forests belonging to the Crown, the Church and the nobility were sold off. Many communal forests were shared out among the inhabitants. The outcome was swift deforestation, because the new owners obtained a higher return by switching their land to agriculture than by keeping their forests.

As the adverse affects of this deforestation (what we would nowadays call the externalities) became clear, the State reforged its forestry policy from the 1820s onwards, initially drawing on the German model. Over the rest of the 19th century there were alternating phases of liberalism (public forests being sold off) and government intervention (reforestation of hill and mountain areas, stabilizing of sand dunes).

After the second world war the lethargy of the government forestry services was strongly criticised, at a time when economic expansion was strong in every area. The National Forestry Fund (FFN) was set up to encourage replanting by private land owners. In addition, the management of public forests was entrusted to a public industrial and commercial entity, the National Forestry Office (ONF). In recent years however, the operation of both bodies has frequently come under criticism from ecologists and the general public, who consider that too much importance is given to replanting with conifers.

A major problem today is the trend towards the reduction in and concentration of agricultural activity. Whatever the future of Europe's Common Agricultural Policy, several million hectares of land are likely to be taken out of farming. What is to be done with them? If they are replanted, the new forests that will spring up on fairly good agricultural land will be more productive than the "older" forests. What will then become of the latter, and how are they to be managed?

That question is already acute in the southern parts of France. In these regions the process of abandoning marginally productive land began in the mid-19th century, and is now far advanced. Land of this type has been taken over by spontaneous woodland, now occupying large areas. Little timber is actually produced. The areas are important for the pattern of life and the landscape, on the other hand, so there is significant externality and non-excludability aspects. Two dangers threaten these wooded areas, fire (the plant formations on them are often highly combustible) and uncontrolled urban development. Around the major towns and in tourist sectors, near the coast or around ski resorts and beauty spots, the value of these areas as building land is far greater than the market value of all the other resources that can be obtained from them. The authorities accordingly have to regulate land use in order to limit the externalities flowing from uncontrolled urban development, as part of their general land management policy.

In the European context forestry policy has long been tied, in one way or another, to strategic considerations of land use and management. Sedjo (Chapter 5) notes that in North America it is reasonable to consider that the production and recreation functions of most forests can best be managed by means of the timber market, or the recreation market, and that only some functions of protection, in fact in fairly small areas, properly belong to the government. In Europe the close interconnections between forest management and general land management mean that it hardly seems possible to replace overall government supervision simply by market forces, without running the risk of an anarchic carve-up of land areas, with serious adverse externalities.

III. TAKING THE VERY LONG TERM INTO ACCOUNT

Conventional foresters in Europe have often been criticised by economists for using few if any economic criteria when making decisions about forestry operations. As Mr. Sedjo noted, back in 1849 Faustman attacked their use of biological rather than economic criteria for determining rotation periods. The fact is that in many European countries, and France in particular, government foresters have long placed regeneration at the forefront of their concerns. Forest exploitation and the harvesting of timber were, so to speak, subsidiary to regeneration. Regenerating forest to the same standard as the forest exploited, or higher, was regarded as an absolute requirement, and even as a moral duty at the very root of forestry ethics.

This constraint can be observed while taking fuller account of economic criteria; in other words it is still possible to optimise subject to certain constraints. At the same time, many forestry decisions (in particular as regards regeneration) involve the very long term, over 50 years into the future. It is well known that the outcome of long-term optimisation will depend very much on the discount rate that is selected. The choice of shadow prices for the resources that will produced in the long term is equally decisive.

The Public Management of Forestry Projects, published by the OECD in 1986, presents a model, based on principles developed by Claude Henry, Director of the Laboratoire d'Econométrie of the Ecole Polytechnique in Paris, which shows how to combine the concepts of discount rates and shadow prices to determine the optimum form of management for a natural resource. In particular, if we adopt shadow prices for natural resources which are rising (at constant prices) in terms of the shadow prices for other goods which the economy produces, biological cycles of production can be justified. In addition, the adoption of rising shadow prices may be justified by the fact that the productivity of natural resource production processes is limited in biological terms, while the productivity of production processes for man-made items seems capable of increasing indefinitely. The relative scarcity of natural resources hence increases, warranting a rise in their shadow prices.

It can thus be shown that, with certain assumptions on future shadow prices, it is not economically inconsistent to adopt biological forms of exploitation.

A specific example of a change in the way in which the long term is taken into account is provided by the forestry policy in a number of Länder in the Federal Republic of Germany, Lower Saxony in particular.

For many years German foresters very largely planted artificial forests of spruce alone. That was the type of forest which, in theory, was most productive and represented the economic optimum. These artificial forests have in fact fallen victim to assaults of many kinds: trees broken by heavy snowfalls, plantations felled by violent wind storms, and recently damage from acid rain. The expected harvest of timber was not achieved in many cases. Over the last few years the foresters have changed policy radically: they are now selecting the species most suited to ecological conditions, and the great majority of plantations are mixed. The forest thus produced is far closer to what a natural forest would be like, and it is hoped it will stand up far better to the assaults of the weather (storms, snow) and pollution (acid rain). It can easily be shown that this policy is economically rational, provided that the risk factor is incorporated in the optimisation models.

To conclude, the choice of an economically optimum form of management for a natural resource depends not only on the discount rate but also on assumptions as to future shadow prices and the level of risk. With different assumptions quite substantially different forms of management can be obtained, all economically optimum given the assumptions. Accordingly, the chief advantage of an economic model is to ensure consistency in the decisions which are based on the model. In addition, many forestry decisions would be more consistent if they took greater account of economic criteria, once the basic assumptions concerning the long term have been set.

IV. TOWARDS CONSENSUS FORESTRY ASSET MANAGEMENT

Private management and government management are today the two main forms of management for natural resources. Yet it should not be forgotten that management has in the past taken other forms as well. For the future, the introduction of new forms of management, negotiated between the social partners, may be a promising way of overcoming the private/government management dichotomy.

To illustrate this point one can refer to a study by Henry Ollagnon, from the French Agriculture Ministry, on the management of natural risks in mountain areas, in the Ubaye valley in the southern French Alps. Right back to the most distant times for which adequate records are available, i.e. the 17th and 18th centuries, highly elaborate procedures were in place to manage natural risks. There was a high risk of flooding or landslides destroying a large proportion of the crops. Very detailed regulations drawn up by the rural communities laid down maximum levels of use for forests and grazing, and obligations to maintain the banks of mountain streams. Local officials were responsible for enforcing the rules; penalties for non-observance were severe.

At the French Revolution these rules of community management were abolished, and the absolute rights of private owners were proclaimed. The result was that many forms of maintenance of the natural environment were abandoned, some former communal lands were cleared of timber, and flooding again became common. In the light of the inability of individual management to control risks, the central authorities decided to intervene and in the 1860s set up a department for mountain land restoration (RTM) responsible for controlling mountain streams and replanting eroded slopes with trees or grass. The service operated principally by expropriating and then replanting forests. Its work, together with the withdrawal of farmers from more marginal land, cut the risks very appreciably. But adverse aspects, what we would term government failures, appeared, while at the same time the nature of the risks was changing dramatically with the progressive abandonment of hill farming and the spread of tourism in the mountains. RTM, whose work was increasingly confined to the land it had earlier expropriated, seemed bound to disappear. Led by farsighted agricultural engineers, it switched roles and became a provider of services and a reliable partner for all the people and agencies (local authorities, ski resorts, etc.) concerned with the control of natural hazards.

This one area has thus seen four successive forms of management of natural hazards: community management as a "bien commun" (which is radically different from the English law concept of common property), private management, government management and now the beginnings of a form of negotiated management.

Thus, with natural hazards, negotiated or consensus management seems to avoid the major defects of both private and government control. It also seems to be a promising way of managing many types of natural resources.

This is because natural resources frequently have multiple uses, of concern to numerous actors. If all their uses could be simply expressed in monetary terms, if there were no externalities or non-exclusion effects, market forces would be the best means of "negotiating" between all these uses and all these actors. That is rarely the case. When alternative uses have no market value, a non-market form of negotiation needs to be established. This will deal in values that are not directly money-related. Trials with the introduction of negotiated management are currently under way in France with water (the river Sèvre in the Nantes region is one example).

When setting up these forms of negotiated management a number of questions have to be tackled:

-- Which actors can legitimately take part in the negotiations?

-- What tangible form should the outcome of the negotiations take: monetary compensation, contractual commitments, etc.?

-- What are the objective data on which the negotiations can be based? It is quite possible that in future sophisticated computer techniques, such as specialised systems using large-scale data bases, will provide very valuable negotiating aids. Expert systems could for example be used to devise scenarios very quickly, evaluate the consequences and present the findings for the negotiators to consider.

This approach is somewhat similar to the environmental mediation that is developing, in the United States in particular, to settle some environmental conflicts.

In fact the management of forestry areas, in Europe at least, shares the following characteristics with the management of many other natural resources:

a) The future of these resources, and their regeneration over the very long term, needs to be taken into account;

b) Management of these resources is tied in with the overall issue of land management;

c) These resources can have multiple uses, not all realised at present;

d) These multiple uses are of concern to multiple social actors;

e) Many of these uses do not have a value determined by the market.

One means of meeting these difficulties is currently being developed in France in the form of asset or heritage management ("gestion patrimoniale" (1). Its main characteristics are concern for the long term, and the importance of forms of management subject to negotiation among the agents concerned.

The basic idea is that neither market forces nor government regulation are capable on their own of giving full expression to the quality objectives held by certain actors. These quality objectives relate to both the current state of the heritage formed by our natural resources and its long-term future. Fresh forms are needed to give expression to these quality objectives and translate them into management on the ground. It could be of great interest to analyse cases, from various OECD countries, where procedures of this type are being implemented, and to make comparisons in order to identify the principles of natural resource management that can avoid the flaws of market management and government management alike.

NOTES

1. Two definitions of the terms heritage or asset ("patrimoine") may be proposed. The first is close to the economic sense: an asset is a good that (with proper management) can keep for the future some potential to adjust to uses that cannot be foreseen at present. The second, developed by Henry Ollagnon, is close to the socio-political sense: an asset is a package of material and intangible elements which combine to safeguard the identity of the holder and ensure its gradual adjustment in a changing world.

Chapter 8

THREE CURRENT LAND USE ISSUES IN THE UNITED STATES:
(A) MANAGEMENT OF THE NATIONAL PARKS, (B) RESTRUCTURING
SOIL CONSERVATION POLICY, (C) LAND DISPOSAL OF TOXIC WASTES

Dr. Douglas D. Southgate

The United States is an excellent arena for analysing how market forces and public policy combine to influence land use decisions. With the exception of problems observed where a large population of poor farmers and herders is concentrated in a fragile natural environment (a pervasive phenomenon in the third world), every major land use conflict arises in the country. Residents of the most crowded nations of Europe or east Asia would recognise the controversies provoked in the most densely populated parts of the United States (e.g. the northeast, California, and Florida) by the impending loss of farmland to commercial or residential development, by the siting of locally undesired enterprises (e.g. nuclear power stations and hazardous waste facilities), or by the potential loss of scenery and related environmental amenities (through changes in agricultural techniques, through land use conversion, etc.). At the other extreme, some parts of Alaska and the region extending westward from the Rocky Mountains to within 200 km of the Pacific Ocean are lightly populated, possess major reserves of minerals and other raw materials, are ecologically sensitive, and offer attractive scenery and recreational sites. Development of such regions involves issues that are very familiar in rural Australia and Canada.

Conflicts over land use in the United States are resolved in a heterogeneous institutional environment. In the western part of the country, the federal government owns large tracts of land. Thus, it must frequently decide whether to use a site for its amenity value or as a source of raw materials. Land use planning elsewhere in the United States is a local prerogative. Some states, counties, and municipalities have instituted controls on the use of privately-owned land similar to those exercised by European governments, while others have adopted a laissez-faire stance toward land markets (Healy and Rosenberg). Finally, as in other countries, decisions made by American land owners are influenced by tax codes, subsidies, environmental regulations, and agricultural policy.

It is impossible to summarise adequately all land use issues now facing the United States in a single chapter. Instead, discussed in the pages that follow are three major issues. They have been selected because similar issues face other OECD Member countries and because the use or restructuring of economic incentives is a central element of many proposals to resolve each of them. In order of discussion, the three issues are: (a) managing national parks so as to preserve threatened eco-systems while still providing adequate recreational opportunities, (b) restructuring soil conservation policy so as to enhance the future productivity of agricultural land and to reduce the downstream costs of soil erosion, and (c) land disposal of toxic wastes.

I. TRADE-OFFS BETWEEN RECREATION AND ECO-SYSTEM PRESERVATION IN UNITED STATES NATIONAL PARKS

Since its inception, in 1916, the United States National Park Service has struggled continuously with the problem of striking a proper balance among competing uses of the land it is charged to manage. The agency's enabling legislation offers only the most general guidance for resolving land use conflicts, the Park Service being instructed "to conserve the scenery and the natural and historic objects and the wildlife therein and to provide the enjoyment of the same." Given this vague mandate, resolution of the trade-off between eco-system preservation and recreation has varied over time as the relative power of the agency's two primary constituency groups --environmentalists and recreators -- has waxed and waned (Foresta).

Also, management decisions have reflected the Park Service's internal politics. The lion's share of the agency's budget and personnel has long been used to service the needs of visitors to national parks. Given this tradition, the Park Service bureaucracy has resisted undertaking ecological research, a critical first step in any programme to preserve unique natural environments. A decade after the United States Secretary of the Interior, who is responsible for the national parks, ordered the agency to place more emphasis on eco-system preservation, generally, and ecological research, specifically, the Conservation Foundation (1972) reported that the Park Service's annual expenditures on natural science research were slightly over US $1 million, less than 1 per cent of the annual operating budget.

Largely because of the low priority assigned to scientific investigation, the Park Service does little more than to protect designated wild areas from the more obvious forms of human encroachment (e.g. resource extraction and modes of recreation that destroy habitat). Not much is done to correct either for past disturbance of eco-systems (e.g. predator control) or for ongoing and subtle disturbances (e.g. the unintentional introduction of non-native species of flora and fauna). The consequences of this approach are becoming more clear. "Protected" areas are in a state of ecological disequilibrium, as evidenced by the past or impending loss of dozens of mammalian populations in fourteen national parks located between the Rocky Mountains and the Pacific Ocean (Newmark).

As criticism from biologists mounts and as environmental groups become more powerful, the Park Service's priorities will change. Among policy instruments that the agency could use to achieve a different trade-off between recreation and eco-system preservation are various types of economic incentives. Historically, they (e.g. entrance charges and rates charged those who stay overnight in park lodges and camp grounds) have covered only part of the costs of a park visit. During the Carter and Reagan administrations, they have been raised in order that recreation be more financially self-supporting. However, if recreation damages eco-systems, user fees should exceed the costs the agency associates with a visit, the "environmental surcharge" serving as a price signal to recreators to limit their visits and revenues raised being used to remedy recreation's adverse impacts on fragile eco-systems.

One challenge for an agency wishing to limit environmental damages in this way, however, is to set an appropriate surcharge. If the surcharge does not equal marginal environmental damages, then the number of recreational visits will not be optimal. But marginal environmental damages, which reflect society's willingness to pay for the existence of species and habitats as well as for the option to use those resources at some time in the future, are very difficult to estimate (Cummings et al.).

The problem of identifying and pricing for non-recreational values of wild areas would not necessarily be addressed by transferring national parks to the private sector. Whether the transfers were permanent or temporary (i.e. whether permanent title in parkland were awarded to a private firm or whether a firm were only given a lease to manage a park for a certain number of years), privatisation would best serve the goal of eco-system preservation when pursuit of that goal did not conflict with recreation. For example, a "club" of duck hunters formed to acquire property rights in a wetland has a vested interest in maintaining the ecological integrity of that habitat. By contrast, a contractor hired to manage a park containing unique and fragile natural environments would find it difficult to charge society for what it is willing to pay to preserve those environments, which implies that it would have little incentive to collect an optimal environmental surcharge from visitors. Thus, a government interested in preserving eco-systems would be obliged to force contractors both to collect such a surcharge and to use the proceeds to remedy the adverse impacts of recreation in sensitive natural environments.

Numerous privately-owned recreational sites are found throughout the United States (and in many other OECD Member countries). That access to practically all of them is rationed on the basis of price suggests that a user fee would be a highly efficient means to help control the number of recreators passing through national parks. Regardless of who might be in charge of collecting fees in the future, though, government will inevitably have to verify that fees are covering the environmental costs of recreation in national parks. If the Park Service retains its managerial responsibilities, bureaucratic review of environmental surcharges will be necessary. Government will likewise have to review surcharges if the private sector assumes management of the parks, since private firms, left to themselves, would not always gain by trying to identify optimal surcharges, by collecting same, or by spending the proceeds on management of eco-systems having little or no recreational value. Determining the relative costs of the bureaucratic process and the review of private surcharge schemes is an empirical issue that should be addressed in future debate over privatisation of the national parks.

II. RESTRUCTURING SOIL CONSERVATION POLICY

For the most part, the United States government's efforts on behalf of fragile terrestrial environments are concentrated on publicly owned land, principally the national parks. Economic incentives have never figured prominently in those efforts. By contrast, economic incentives have long been used to promote conservation of the country's farmland, which is almost entirely privately owned. During most of the 50-year history of the Agricultural Conservation Program (ACP), measures to arrest soil erosion (which can be broadly defined to include soil salinisation and other adverse changes in soil quality in addition to the removal of soil from a parcel of agricultural land by wind or precipitation) have been "cost-shared" (i.e. subsidised). Farmers have also been paid to withdraw land from production in order to reduce soil erosion.

For more than a decade, the ACP has been criticised in the United States. Observers doubt that ACP incentives are being targeted in areas where soil erosion either seriously threatens land productivity or creates major downstream costs. Also, ACP economic inducements to participate in conservation programmes are alleged to have been too generous. To understand why United States soil conservation policy has evolved to the stage where such criticisms are valid, the origins of that policy are noted early in this section. Then, options for restructuring economic incentives to promote soil conservation more effectively are examined.

The ACP began less than two months after the Roosevelt administration's initial effort to control food and fiber production and to regulate commodity markets, the Agricultural Adjustment Act of 1933, was ruled unconstitutional by the United States Supreme Court, in January of 1936. Setting the pattern for future initiatives ostensibly undertaken to conserve land resources, the ACP offered payments to farmers who switched from soil-depleting crops (i.e. intensively cultivated row crops like maize and soybeans) to soil-conserving crops (e.g. grasses and legumes). Cost-sharing was also made available to farmers who adopted "soil-building" conservation practices. These initiatives substituted for the 1933 Act inasmuch as they were designed and executed in a manner consistent with the achievement of two basic objectives the Roosevelt administration (and subsequent governments) set for agricultural policy: enhancement of farm income and lessening of production capacity (Rasmussen).

A critical feature of United States soil conservation policy since the 1930s has been that locally elected committees composed almost exclusively of beneficiaries of the ACP and of technical assistance programmes designed by the Soil Conservation Service (SCS), have executed policy. These local representatives are potent advocates for the status quo in soil conservation policy. Largely through their efforts, federal allocations to erosion control programmes have stayed constant, in real terms, since the late 1940s (Easter and Cotner). Moreover, Congress, lobbied by local committees, has resisted efforts made by every successor to Roosevelt to "rationalise" soil conservation policy, for example by combining SCS and ACP programmes in a single agency (Rasmussen).

Farmers' influence on soil conservation policy can be inferred from criticisms leveled at that policy during the past ten years. A decade ago, the United States Comptroller General (1977) presented evidence that a substantial amount of cost-sharing funds was being disbursed to pay for practices that enhanced crop production without greatly reducing soil loss. The same agency has also observed that funds were distributed so as to achieve geographical equity (and implicitly, keep all local constituencies satisfied) rather than being targeted on areas where land productivity was seriously threatened by soil erosion or where soil erosion severely impaired water quality (United States Comptroller General, 1983). In addition, there has been considerable "slippage" in commitments made by farmers under the ACP and related programmes to reduce hectarage planted to erosive row crops (Ericksen and Collings), which is hardly surprising inasmuch as the local committees monitor compliance with hectarage reduction commitments.

Responding to these criticisms, the United States Congress, through the Soil and Water Resources Conservation Act of 1977, directed the United States Department of Agriculture to assess the nation's soil and water resources and to develop a more effective conservation programme. One probable long-term result of this review is that ACP funding of production activities that are profitable for farmers while having little impact on soil erosion will probably be reduced. In addition, an effort to target programmes more on "problem areas" will be made.

Beyond accepting targeting as a general guideline, however, SCS and other agencies find it difficult to operationalise that principle. As Clark et al (1985), Crosson and Stout (1983), and Lovejoy and Napier (1986) observe, an agency wishing to target its soil conservation programmes must have information about the costs associated with on-farm as well as off-farm impacts of soil erosion (or reasonable proxies for those costs) along with information about the relative contribution of erosion from agricultural land to downstream water quality problems.

In addition, the criteria for selecting targeted areas are almost always the subject of debate. Traditionally, the stated purpose of United States soil conservation programmes has been to maintain land productivity, although Crosson and Stout (1983) stress that purpose has been served poorly by United States soil conservation policy. Since passage of the 1972 Amendments to the Federal Water Pollution Control Act, though, a mandate for reducing non-point source pollution from agricultural land has also existed. Unfortunately, because areas where soil erosion greatly affects productivity do not coincide closely with areas where the downstream costs of soil erosion are high (Ribaudo), targeting decisions are bound to be heavily influenced by the relative importance decision-makers attach to downstream impacts versus on-farm productivity effects.

Furthermore, regardless of how well (or poorly) soil conservation policy is targeted on geographical problem areas, government faces the task of selecting a set of instruments to induce farmers' cooperation in that policy. Discussed in the rest of this section is how well four economic incentives for erosion control -- (a) subsidies, (b) taxes, (c) land retirement bids, and (d) cross compliance -- could be used or restructured to serve the latter purpose.

Subsidising Erosion Control

As already indicated, subsidies, long extended to American farmers who adopt erosion control measures, have rarely reflected the social value of reduced soil loss. Cost-share rates have been set quite high (typically, far above 50 per cent), even for practices applied to land where erosion does no great damage to downstream water quality and even when farmers would have been willing to self-finance those practices to a considerably greater degree. Furthermore, cost-share funds have often been allocated on a "first-come-first-served" basis by local committees more interested in maximising the number of farmers participating in the ACP than in dealing with particularly severe erosion problems (Easter and Cotner).

A reform that would partially reduce the inefficiencies inherent in this regime would be to link subsidy rates to erosion rates. This would, however, be an imperfect solution since erosion's impacts on land productivity can vary considerably from place to place (Crosson and Stout). In addition, varying subsidy rates solely on the basis of estimated soil loss rates takes no account of the downstream benefits of controlling erosion from any given parcel of land. To estimate the latter, one needs to know how much of the soil washed from the parcel ends up in a lake, stream, or other water body threatened by sedimentation, eutrophication, or some other impairment of water quality. In addition, other sources of pollutants must be ascertained, as should water users' willingness to pay for improved water quality. None of this information, all necessary for targeting erosion control measures where they will have the greatest possible impact on downstream water quality, is easy to obtain.

Taxing Soil Loss

Baumol and Oates (1975) describe some general principles to be followed when implementing a Pigouvian tax on natural resource depletion: if the marginal social costs of depletion, demand and supply conditions facing natural resource users, and market imperfections elsewhere in an economy are known, then an agency can identify socially optimal resource use along with a tax for inducing that state. The catalogue of information needed for execution of Pigouvian taxes on agricultural run-off and other forms of non-point source pollution is essentially the same as the data requirements for socially optimal subsidisation of soil conservation.

Addressing soil erosion problems with a Pigouvian tax is further impeded by its impacts on income distribution. Whereas a subsidy scheme enhances (or at least does not reduce) farm income, a tax on soil loss would accomplish exactly the opposite. Some available research indicates that the primary consequence of imposing such a tax might be to reduce farm income. For example, Jacobs and Casler (1979) estimated that a tax of $100 assessed on each kilogram of soluble phosphorus run-off (a major non-point source pollutant) would cause farmers in a small watershed located in New York state to decrease that form of pollution only by 20 per cent. Such a tax would cause farmers to spend a little more than $100,000 on run-off control while generating nearly $1 million in revenues for the government. Of course, farmers' choices between paying the tax and paying for erosion control would be influenced by the tax rate. The two investigators found that a $200/kg tax would cause farmers to spend $600,000 controlling soluble phosphorous run-off, thereby reducing that form of pollution by 50 per cent. At the higher tax rate, revenues would exceed $1 million (Jacobs and Casler). (These estimated income and behavioural impacts probably reflect the structure of the linear programming model used in Jacobs and Castler's research, however.)

Bidding over Land Retirement

Since the inception of the ACP, the United States government has contracted with farmers to withdraw land from production. Until recently, there was little negotiation between government and farmers over compensation paid the latter to withdraw land from production. Since passage of the 1985 Food Security Act, however, farmers have been bidding to place highly erodible land in a "Conservation Reserve." Those willing to accept the least amount of compensation to reduce acreage planted to erosive crops end up participating (Dicks).

Of course, farmers' willingness to accept compensation for retiring cropland depends on the net returns that could be earned by farming it. Those net returns are enhanced by agricultural policies, like those pursued by OECD Member countries, that maintain commodity prices at artificially high levels. In addition, bidding by potential participants in the Conservation Reserve Program (CRP) is influenced by the option to participate in another of the government's land withdrawal efforts: the Acreage Reserve Program (ARP). Farmers enroll in the ARP by setting aside 20 per cent of the "base area" planted to crops, like maize and cotton, the prices of which are supported by government. They then become eligible to receive support prices for whatever they produce on the rest of their land.

Taff and Runge (1986) point out that cropland placed in the Acreage Reserve for a growing season is the least productive (and usually the most erodible) land. CRP bids, then, reflect the higher opportunity cost of retiring land having higher yields. In addition, they point out that CRP bids are further inflated because a farmer's ARP base is automatically reduced for 10 years as she or he places a hectare in the Conservation Reserve. Analysing data on CRP bids from several parts of the state of Minnesota, Taff and Runge (1986) estimate that these two conflicts between the CRP and ARP cause average CRP bids to be about 20 per cent higher than they would otherwise be. They propose eliminating the conflict between the two programmes (and thereby reducing farmers' CRP bids) by making only highly productive and relatively non-erodible eligible for the ARP and by making lands eligible for the CRP (i.e. highly erodible land) ineligible for the ARP (Taff and Runge).

Cross Compliance

Cross compliance is a fiscal mechanism for enforcing conservation regulations. All farmers receiving payments under agricultural programmes would be obliged to submit erosion control plans for government approval. Failure to do so or failure to abide by an approved plan would jeopardise programme payments.

A major cost to the government of such a scheme would comprise expenses of approving and monitoring farm-level conservation plans. These expenses could be large. In addition, there is reason to suspect that "cross-compliance" would be a poorly targeted policy instrument (Cook). Many farmers on land where soil erosion is high do not receive agricultural programme payments. Reichelderfer (1985) has found that only a third of the land that is being cropped in spite of its high erodibility is used by participants in United States farm programmes. Cross compliance, then, would not influence the behaviour of those who use the remaining two thirds of that land. In addition, withholding federal payments from farmers who do not conserve soil would not be an effective policy instrument during times, like the early 1970s, when strong demand for agricultural commodities causes those payments to dwindle away to zero. Finally, like a tax on soil loss, the primary impact of attempting to enforce cross-compliance might be to reduce farm income rather than to reduce soil loss. For example, participants in United States farm programmes who use highly erodible land might continue to find planting erosive crops the most remunerative land use option in spite of cross compliance rules. They would therefore do so, accepting the loss of federal payments.

Conclusions

The debate over economic incentives to promote soil conservation is bound to continue for years. The technical challenge of designing an optimal tax or subsidy is considerable, as noted above. Similarly, the costs to government of adopting Taff and Runge's (1986) proposal for reducing bids on land enrolled in the Conservation Reserve (i.e. categorising all farmland according to suitability for the CRP or ARP) would probably be large. Furthermore, reforming soil conservation policy is politically contentious, many farm groups favouring the continued use of the ACP and other programmes to transfer income rather than simply to influence natural resource management.

The debate over soil conservation policy is bound to become more heated as a broader perspective is taken on the market forces and governmental actions affecting agricultural use of land. Writing soon after the commodity price boom of the early 1970s, for example, Timmons and Cory (1978) speculated that a similar resurgence in the value of crops would cause soil erosion to rise by 70 per cent in the midwestern United States. Farther to the west, high prices (sustained by strong demand for United States grain as well as public policy) have caused relatively non-productive land that is highly susceptible to wind erosion to be cultivated (Heimlich). Any attempt by government to ameliorate the resource use impacts of high prices (e.g. through subsidies) would probably be costly. Alternatively, any effort to contain peak prices (in order to avert natural resource depletion and to accomplish other goals) always arouses opposition.

III. LAND DISPOSAL OF HAZARDOUS WASTES

Federal involvement in the management of wilderness lands has a long history in the United States. The first national park, Yellowstone, was founded in 1872. Also, national soil conservation policy dates from the 1930s. By contrast, land disposal of hazardous wastes was generally treated as a matter of state, rather than federal, concern until passage of the Resource Conservation and Recovery Act (RCRA) of 1976.

That legislation was passed in response to growing awareness of the serious threats both to human health and to the environment posed by the improper treatment, transportation, and disposal of hazardous wastes. In addition, passage of the RCRA reflected dissatisfaction with economic sanctions traditionally used to discourage inadequate handling of toxic residuals. In most American states before enactment of the RCRA and complementary laws and regulations, those injured by toxic substances leaking from landfills and other sites had little recourse but to sue for damages, as provided for by common (or, court-made) law.

McAllister (1982) indicates that, in and of itself, common law is an inadequate mechanism for dealing with land disposal problems. Plaintiffs lack standing to sue careless landfill operators until they have suffered some sort of injury. Also, the probability that a negligent party will pay the full cost of a leak of dangerous materials is considerably less than 100 per cent because the link between that party's behaviour and injury to others is frequently difficult to prove (which implies that litigation costs in such cases are often quite high) and because that link sometimes becomes apparent only after the passage of many years (by which time the offending party might be bankrupt or difficult to locate). For these reasons, the common law does not always provide sufficient inducement to reduce environmental risks.

Largely because of the inadequacies of the common law, a strong emphasis was placed in the RCRA on regulation. In particular, that legislation created a mandate for the United States Environmental Protection Agency (EPA) to enforce safety standards in the handling of a large number of toxic residuals from the time they are generated until their final disposal. Like exclusive reliance on the common law, however, this approach had its shortcomings. Close monitoring of treatment, transportation, and disposal being very costly, it was inevitable that EPA enforcement of RCRA-mandated regulations would be less than perfect. Consequently, the behaviour of firms involved in hazardous waste management, including those in charge of land disposal sites, was not socially optimal.

Theoretical insights into the problems of hazardous waste policy, both before and after passage of the RCRA, are provided by Shavell (1984). As he indicates, the incentives offered by common law are "diluted" by the chance that negligent parties will not be sued. Similarly, information costs prevent a regulatory agency from assuring that the private sector is doing enough to avoid accidents. He argues that an institutional regime featuring both common law and regulatory elements furnishes superior incentives to avoid risky behaviour.

Increasing familiarity with the limitations of the regulatory approach mandated by the original RCRA has led to development of the type of combined regime discussed by Shavell (1984). In effect, the EPA now generates information that plaintiffs can use in lawsuits brought against negligent operators of land disposal sites. Also making the common law a more effective mechanism for dealing with hazardous waste pollution are new rules that enable plaintiffs to "discover" information about defendants' waste management decisions, which facilitates the legal task of proving the latter group's negligence (McAllister).

Finally, even a combination of regulation and common law is inadequate for dealing with all the country's hazardous waste problems. To cover the expense of dealing with consequences of past negligence, a "Super Fund," administered by the EPA, has been established. In the future, it will be used more to cover the costs of negligent behaviour of financially insolvent waste handlers. Of course, insurance schemes, like the Super Fund, do not always provide sufficient economic incentives to reduce environmental risk. If general tax revenues (collected from society as a whole) are paid into the fund, then industry will pollute excessively. Similarly, an individual firm's risk avoidance will be sub-optimal if its insurance payments do not reflect its waste management decisions. In either case, subsidisation of one firm's or one industry's negligent behaviour results in excessive environmental risk (Calabresi).

IV. SUMMARY AND CONCLUSIONS

Much of the literature on natural resource economics deals with the properties of optimal prices charged for goods and services drawn from the environment. Because those prices should reflect all future costs associated with present natural resource use (which often carries economically irreversible consequences), those economic incentives are always subject to challenge. Certainly, this is true of the incentives discussed in this chapter. As indicated above, it is not easy to identify optimal user fees in national parks, which should reflect the marginal environmental damage of recreation. Similarly, it would be difficult to determine an optimal tax on soil erosion or an optimal sanction on leaks from sites where hazardous wastes are stored.

Even when the empirical issues associated with identification of an optimal price signal for natural resource use are settled, an agency would have to confront the vexing political problem of imposing it on the marketplace. This problem arises in numerous contexts. If private firms assume greater managerial responsibility for United States national parks, for example, the Park Service will have to monitor the environmental surcharges added to recreational user fees. Similarly, a scheme to tax soil loss or a proposal to target erosion control subsidies better would arouse the opposition of those who prefer the present arrangement of using soil conservation policy as a mechanism to transfer income to farmers.

Finally, much economic literature emphasizes that economic incentives are superior to regulation. Certainly, this general observation applies to United States soil conservation policy, the probable cost of attempting to set and to enforce conservation standards for millions of hectares of cropland being quite high. But in other contexts, as Baumol and Oates (1975) observe, the public interest is not served by forcing policy makers to choose between economic incentives and regulations. Parks are sometimes managed better, for example, when recreational access is limited both by user fees and by regulation. Similarly, as McAllister (1982) and Shavell (1984) indicate, regulation and the common law are complementary policy tools for dealing with hazardous wastes management problems. The policy choice, then, is not always regulation versus economic incentives. Rather, policy makers must often identify ways to make a combined regime more effective.

REFERENCES

1. Baumol, William J. and Wallace E. Oates. The Theory of Environmental Policy. Englewood Cliffs, New Jersey: Prentice Hall, Inc., 1975.

2. Calabresi, Guido. The Costs of Accidents. New Haven: Yale University Press, 1970.

3. Clark, Edwin H., Jennifer A. Haverkamp, and William Chapman. Eroding Soils: The Off-Farm Impacts. Washington, D.C.: The Conservation Foundation, 1985.

4. Conservation Foundation. National Parks for the Future. Washington, D.C.: 1972.

5. Cook, Ken. "Cross-Compliance: Is It Bold, Menacing, or Just Plain Dumb?" Journal of Soil and Water Conservation 39:4(1984)250-251.

6. Crosson, Pierre R. with Anthony T. Stout. Productivity Effects of Cropland Erosion in the United States. Baltimore, Maryland: Johns Hopkins University Press, 1983.

7. Cummings, Ronald G., David S. Brookshire, and William D. Schulze (ed.). Valuing Environmental Goods. Totowa, N.J.: Rowman and Allanheld, 1986.

8. Dicks, Michael R. "Conservation Reserve Signup Mirrors Successful Soil Bank Program," Agricultural Outlook (September, 1986) 30-32.

9. Easter, K. William and Melvin L. Cotner, "Evaluation of Current Soil Conservation Strategies," in Harold G. Halcrow, Earl O. Heady, and Melvin L. Cotner (eds.) Soil Conservation Policies, Institutions, and Incentives. Ankeny, Iowa: Soil Conservation Society of America, 1982.

10. Ericksen, Milton H. and Keith Collins. "Effectiveness of Acreage Reduction Programs" in U.S. Department of Agriculture, Economics Research Service, Agricultural-Food Policy Review (agricultural economics report 530). Washington, D.C.: U.S. Government Printing Office, 1985.

11. Foresta, Ronald A. America's National Parks and Their Keepers. Baltimore: Johns Hopkins University Press, 1984.

12. Healy, Robert G. and John S. Rosenberg. Land Use and the States (2nd ed.). Baltimore, Maryland: Johns Hopkins University Press, 1979.

13. Heimlich, Ralph. "Sodbusting: Land Use Change and Farm Programs," Land Economics 62:2 (1986) 174-181.

14. Jacobs, James L. and George L. Casler. "Internalising Externalities of Phosphorus Discharges from Crop Production to Surface Water: Effluent Taxes versus Uniform Reductions," American Journal of Agricultural Economics 61:2(1979)309-312.

15. Lovejoy, Stephen B. and Ted L. Napier (eds.), Conserving Soil: Insights from Socio-economic Research. Ankeny, Iowa: Soil Conservation Society of America, 1986.

16. McAllister, Kevin. "Hazardous Waste," in Michael S. Baram (ed.). Alternatives to Regulation: Managing Risks to Health, Safety, and the Environment. Lexington, Mass.: D.C. Heath and Company. 1982.

17. Newmark, William D. "A Land Bridge Island Perspective on Mammalian Extinctions in Western North American Parks," Nature 325 (29 January, 1987) 430-432.

18. Rasmussen, Wayne D. "History of Soil Conservation, Institutions and Incentives," in Harold G. Halcrow, Earl O. Heady, and Melvin L. Cotner (eds.), Soil Conservation Policies, Institutions, and Incentives. Ankeny, Iowa: Soil Conservation Society of America, 1982.

19. Reichelderfer, Katherine H. "Do USDA Farm Program Participants Contribute to Soil Erosion?" (agricultural economics report 532). Washington, D.C.: U.S. Department of Agriculture, Economic Research Service, 1985.

20. Ribaudo, Marc O. "Targeting Soil Conservation Programs," Land Economics 62:4(1986)000-000.

21. Shavell, Steven. "A Model of the Optimal Use of Liability and Safety Regulation," Rand Journal of Economics 15(1984) 271-280.

22. Taff, Steve and C. Ford Runge. "Supply Control, Conservation, and Budget Restraint" (Dept. of Agricultural and Applied Economics staff paper P 86-33). St. Paul, Minnesota: University of Minnesota, 1986.

23. Timmons, John F. and Dennis C. Cory. "Responsiveness of Soil Erosion Losses in the Corn Belt to Increased Demands for Agricultural Products," Journal of Soil and Water Conservation 33:5(1978)221-226.

24. U.S. Comptroller General. To Protect Tomorrow's Food Supply, Soil Conservation Needs Priority Attention (report to Congress, CED 77-30). Washington, D.C.: U.S. General Accounting Office, 1977.

25. U.S. Comptroller General. Agriculture's Soil Conservation Programs Miss Full Potential in the Fight against Soil Erosion (report to Congress, GAO/RCED-84-48). Washington, D.C.: U.S. General Accounting Office, 1983.

Chapter 9

TOWARDS IMPROVED LAND MANAGEMENT

Dr. Heino von Meyer

I. LAND MANAGEMENT: A DEFINITION

Better management of the natural resource in question requires that the concept "land" be understood not only as surface area, but also as the soil eco-system.

Management of land area is concerned with the spatial distribution of alternative forms of land use. Management of the soil eco-system aims at the conservation and improvement of its properties and characteristics in respect of various specific functions. Here we are especially concerned with both the ecological regulative function, and the agricultural productive function.

II. THE MAIN TASKS OF LAND MANAGEMENT

With regard to the use of land area the search for improved land management especially involves the future development of:

-- the settlement structure, i.e. the growth and distribution of residential, commercial and transport areas;

-- nature conservation areas, i.e. nature reserves or other areas not or only extensively utilised; and

-- agricultural and forestry areas, i.e. size and cultivation structures according to regional differentiation.

The future of the countryside and the aesthetic and ecological diversity on which the quality of life in urban as well as rural areas depends will be decisively affected by such developments.

With regard to the management of the soil eco-system it will be essential to confront the following problems:

-- soil structure damage due to wind and water erosion as well as due to compaction; and

-- soil pollution damage caused by agricultural production (mineral and organic fertilizers, pesticides, irrigation) as well as by air pollution (acid rain, heavy metals and radioactivity), waste deposits, etc.

III. DIFFERENT PROBLEMS AND PERCEPTIONS

The problems of land management and the way they are perceived by the public and policy-makers in the different OECD countries vary considerably. The main reasons for this are not only the differences in natural conditions but also the differences in institutional traditions which are reflected in legal, socio-economic and cultural structures.

It is no wonder that in countries which have been densely populated for centuries and today have a population density of 250 inhabitants per sq km or more (West Germany: 246, Netherlands: 350) there is a very different concept of "land management" from that found in "immigration countries" such as the United States or Australia which have population densities of 25 inhabitants per sq km and less. However, even within Europe there are large differences in the way the problems are perceived, e.g. between the northern and southern regions of the European Economic Community.

Generally it would appear that the main difference is due to the fact that in areas with low population densities conflicts between different types of land use could for a long time be relatively easily solved by a spatial division of the different functions of land use. In more densely populated areas, on the other hand, an integrated management of different overlapping aims and functions had to be implemented a long time ago.

Such details ought to make it clearer why the main concern of "land management" in the United States is the control and prevention of soil erosion (e.g. USDA, 1982), whereas in Europe other questions have traditionally received more attention (BOUWENS/DOUW, 1986; SRU, 1985; DLR, 1986), such as, for example:

-- the prevention of uncontrolled urban growth; and

-- the prevention of soil pollution.

In the Federal Republic of Germany these concerns have led to the formulation of a concept known as "Bodenschutzkonzeption" -- land conservation programme" -- (BMI, 1984) which is essentially focussed on three phenomena:

-- "Waldesterben" (dying woodland);

-- nitrates (especially in drinking water); and

-- urban spread.

IV. GROWTH OF SETTLEMENT AREA

In the Federal Republic of Germany it is considered problematic (IMAB, 1984) that settlement areas (built-up areas for buildings, infrastructure and transport) have increased in the last three decades.

	(million hectare)		(% of the total land area)	
-by	1.0	or	4.3	
-from	1.9	or	7.5	in 1950
-to	2.9	or	11.8	in 1981.

From 1965 to 1980 the settlement area increased by 20 per cent, while population only increased by 5 per cent, and employment even decreased by 5 per cent.

However, this should not be considered as the central question of land management. Improved forms of land management require a better regional distribution of the investment in infrastructure, housing and industry. A more decentralised settlement structure with lower concentrations of built-up areas is probably less harmful to the environment (e.g. less pollution, etc.) even if the total area of land required for such purposes is slightly larger (PRIEBE, 1982).

V. LOSS OF NATURAL AREAS

With very few exceptions almost all European landscapes have been used intensively for thousands of years. As a result there are almost no completely natural areas left. European landscapes are "cultural" landscapes and not "natural" landscapes (v. MEYER, 1984 and 1985). Even today's conservation areas are often "man-made". In order to be maintained they have to be managed. Without extensive forms of grazing, for example, many protected orchid meadows and heath landscapes (e.g. Lüneburger Heide) would disappear and in the long term again become woodland.

In the Federal Republic of Germany conservation areas and national parks scarcely make up 2 per cent of the total land area (BMELF, 1986; p. 355): over 30 per cent of such areas are smaller than 10 ha; and only 17 per cent are larger than 100 ha.

In the Federal Republic of Germany the "Red List" of endangered species shows that: 28 per cent of flowering plants and ferns; 47 per cent of mammals, and 38 per cent of birds are either "extinct" or "endangered at the present time" (StJB, 1985, p. 579).

Numerous studies (SUKOPP, 1981; SRU, 1985) have shown that agriculture has born the main responsibility for the increasing loss of wild animal and plant species in recent years. Thus, for example, over 70 per cent of the vascular plants on the "Red List" are primarily endangered by agriculutral land use. The main causes of this loss of species were identified as the elimination of habitats; drainage, and abandonment of or changes in -- mainly intensification of -- agricultural land use.

Thus, conservation and the better management of natural resources is also dependent on the management of agricultural land.

VI. AGRICULTURAL LAND USE AND STRUCTURAL CHANGE

In the European Community agricultural land use has been subject to a fundamental structural change during the past three decades. This transformation can be characterised (de HAEN, 1985; v. MEYER, 1983):

-- at the farm level by concentration, intensification and specialisation; and

-- at the regional level by a polarization which is characterised by increasing inter-regional differentiation and, at the same time, intra-regional standardization.

These tendencies are both effect and cause of technological and socio-economic changes which have not only affected agriculture but also the economy and society as a whole. While they have not only been determined by policy, there can be no doubt that the scale and direction of this structural transformation was primarily influenced by economic incentives introduced by policy-makers.

If agriculture and many rural regions in Europe are today confronted with a cumulative economic, social and ecological crisis, then it is also because for decades agricultural policy has sent the wrong signals. This transformation of agricultural and rural structures, which can thus be seen to have been falsely oriented, endangers not only the economic equilibrium, but also the ecological balance. With regard to land use this has consequences for the management of land area as well that of the soil eco-system.

VII. AGRICULTURE AND THE MANAGEMENT OF LAND AREA

It is difficult to assess whether or not the agriculural area in use (AA) is larger as a result of the protectionist Common Agricultural Policy (CAP) than it would have been under a more liberal, market-oriented system (CHESHIRE, 1985). There is no doubt that high prices have encouraged the maintenance of production in marginal areas. However, high and increasing prices have also induced technological progress and thereby possibly increased the degree of intensity to such an extent that this may have neutralised or even overcompensated this effect.

There may also be argument about whether the proportions of agricultural land under arable use and used as grassland would be different today -- a question which is not only of interest with regard to the erosion problem. There is, however, some evidence for the fact that animal products, for example milk, are more protected than products such as cereals (HODGE, 1986). This tends to increase the proportion of grassland and this may have had a positive effect on the environment. The fact that grassland was ploughed up in some areas of the EEC after the introduction of milk quotas can also be seen as an indication of this.

However, what is most important is the fact that the variety of small pieces of land in use has been drastically reduced and as a result there has been a reduction in the length of the ecologically active tracts of border land and thus also in many of the elements (boundary ridges, hedgerows, etc.) which used to shape our landscapes (HOUSE OF LORDS, 1984). This loss of diversity has not only been caused by the concentration of farms and parcels of land which was often stimulated by farmland consolidation programmes, but a role has also been played by the increased specialisation seen in reduced crop rotation and the separation of cropping from livestock production, resulting from distorted price relations and increased price stability.

VIII. FUTURE TRENDS IN AGRICULTURAL LAND AREA USE

In view of the present situation and emerging future trends in agricultural markets inside and outside the EEC, it would appear possible that approximately 20 per cent of the agricultural area of the Community could be taken out of agricultural production in the next 15 years without compromising Europe's self-sufficiency objectives (LEE, 1985; COUNTRYSIDE COMMISSION, 1987).

The "set-aside" and "reconversion" of agricultural land is at present being discussed in detail in the Federal Republic of Germany and within the EEC (POTTER, 1985; BOWERS, 1987). As long as the solving of the problems of the agricultural market remain the sole concern here, a worsening of the general status of the "land" resource as defined above would appear to be more likely than an improvement. The "set-aside" or afforestation of areas in many less-favoured regions (especially hill and mountain areas) provides no clear environmental advantages and the burden on the soil eco-system in favoured regions would increase even further as a result of the further intensification of crop and animal production systems.

In order to achieve nature conservation objectives it would appear to be worthwhile to aim at creating an interlinked biotope network for the protection of threatened animal and plant species which ought to cover about 10-15 per cent of total land area (SRU, 1986). This, however, requires a regionally differentiated approach (v. MEYER, 1985).

-- In favourable agricultural regions conservation would require the renewed establishment and reconstruction of biotope structures through the partial contraction and extensification of agricultural production.

-- In less favoured regions, on the other hand, conservation is more a matter of maintaining and developing existing structures and thus also current agricultural production practices.

This already indicates that there is not always a conflict between agricultural production and environmental protection. On the contrary, for many less favoured regions in Europe there is quite often a positive accord between the promotion of regional development and the conservation of the natural environment.

IX. AGRICULTURE AND THE MANAGEMENT OF THE SOIL ECO-SYSTEM

Future trends in agricultural land use within the EEC urgently require greater attention: not only because it would seem possible that within a few years agricultural production will cease in an area which is twice the size of the total increase in settlement area over the past three decades; but also because the productive capability of the soil eco-system will depend decisively on the future organisation of agricultural land use.

Agriculture influences the way the soil eco-system functions: through changes in soil structure; and also through the introduction of certain substances into the soil.

These cannot in themselves be regarded as problems since such changes and practices are essential for agricultural production. They only become a problem when so much emphasis is given to the agricultural productive functions of the soil that other functions, such as the ecological regulative function, become permanently disturbed (OECD, 1986).

X. SOIL STRUCTURE DAMAGE

Damage to the soil structure by wind and water erosion, and compaction is primarily caused by crop production. The structural change in European agriculture outlined above has led to increased problems in this area. The concentration of parcels of land into ever larger fields (land consolidation), together with the removal of the hedgerows or terraces (e.g. in vineyards) that protect against wind or run-off has undoubtedly also increased the risk of soil erosion. The intensification of mechanical tillage has also aggravated erosion problems. However, although soil loss has generally increased, in Europe the topic of erosion has up to now only found limited regional interest, especially in the south.

In addition to erosion, soil compaction has also increasingly become a serious threat in many areas. This is partly the result of the use of heavier machines, partly the result of specialisation with reduced crop rotation and also partly due to the separation of livestock and crop production systems.

Among other things methods of reduced tillage are currently being discussed as a means of avoiding soil structure damage. A critical point, however, is the fact that this would generally require an intensified use of herbicides, which in turn poses new and different risks for the soil and other eco-systems.

XI. SOIL POLLUTION DAMAGE

The damage which can be caused to the soil eco-system by pollution would appear to be even more significant. Agriculture is not the only polluter here and indeed it is often the victim (e.g. acid rain). Nevertheless, it plays an important part through its methods of fertilizing and pest control.

In the course of the above-mentioned structural change the intensity of fertilization and the use of pesticides has increased enormously (v. MEYER, 1983). The consequences of the increased use of nitrogen fertilizer, especially the pollution of ground and drinking water, have now become a general topic of discussion in the field of environmental policy (OECD, 1986). Changes in the ecology of the soil are also linked to these changes in water quality. For example, in some areas the soil's denitrification potential has been exhausted and as a result an important buffer function of the soil has beeen destroyed. The increased use of mineral fertilizers does not bear the sole responsibility for this. The dispersal of liquid manure, particularly in regions with intensive livestock production, would seem to be a much more important factor. This practice is more often seen as waste disposal than fertilization.

The dispersal of large quantities of liquid manure does not just present problems arising from the leaching of nitrate. The foodstuff in intensive pig production also contains heavy metals (e.g. copper), which accumulate in the long term and can impair plant growth. The same is also true of phosphate fertilizers which can also lead to a concentration of heavy metals, especially cadmium (SRU, 1985).

Moreover, some farmers use sewage sludge from water treatment plants as fertilizer. The high concentration of pollutants, particularly of heavy metals, in such sludge has also proved to be a problem here. In part restrictions on land use have now become unavoidable (DLR, 1986).

Irrigation is another agricultural practice which can seriously affect the functions of the soil. In the new member states of the EEC, particularly Greece, Spain and Portugal, it will be especially important that care is taken to ensure that the incentives to irrigate additional areas, which have come with membership of the EEC, do not lead to the salinisation of the soil.

In addition to fertilizer pollution, the increasing use of highly effective pesticides has also had an effect on the soil eco-system. This problem has hardly been researched, but it is quite possible that it could have significant lasting effects.

However, as a last point it must be stressed that agriculture also makes a very positive contribution to the conservation and strengthening of the eco-system's capabilities. This can be seen in the way that agriculutre attempts to counteract the negative effects of non-agricultural pollution (e.g. acid rain) by using fertilizers containing lime. One indication of the extent of soil damage caused by emissions from non-agricultural sources is the rapidly increasing scale of forest damage. According to the most recent survey of forest damage carried out in 1986, 54 per cent of the woodland area of the Federal Republic of Germany is damaged, one third of it (19 per cent of the total) seriously or very seriously. The main cause is considered to be air pollution, in particular sulphur dioxide and nitrogen oxide (BMELF, 1986).

XII. ECONOMIC INCENTIVES: INSTRUMENTS AND INSTITUTIONS

The discussion about the role of economic incentives in achieving better resource management first required clarification of the question as to whether the hitherto unsatisfactory use of resources is to be attributed to the fact:

-- that the effectiveness of the market process is disturbed by the inappropriate use of policy instruments, e.g. taxes and subsidies; or

-- that the market process - even without interference - is not at all in a position to allocate resources efficiently, either as a result of an unsuitable institutional framework, or because of the fact that this is fundamentally impossible.

Here a distinction is often made between "government failure" and "market failure". This only makes sense, however, if the status quo of the institutional framework is considered unchangeable.

If the optimal design of institutions - understood here in the widest sense as the establishment of rights and administrative structures, etc., - is considered a political task, then in both cases it is a matter of "policy failure" either in the application and choice of economic instruments; or in the design of adequate institutions.

In my opinion, the improvement of land management with the aid of economic incentives requires correction at both (v. MEYER, 1987, c). As already stressed a number of times, improved land management is a matter of the optimal co-ordination of the different forms of land use also involving the consideration of long-term precautionary aspects as well as a matter of protecting the soil eco-system, especially against pollution by reducing it at its source.

Both these management tasks require a regionally differentiated approach.

XIII. INSTRUMENTS: THE CORRECTION OF INAPPROPRIATE INCENTIVES

Even within a given institutional status quo there are generally many ways of using economic incentives to improve land management. The difference is often between "first best" and "second best" solutions.

I would like to illustrate this problem by using the example of agricultural policy and agricultural land management. This example also has relevance for other fields and, for instance, can be applied to regional investment aids and the distribution of industrial and residential areas. It is also relevant for transport tariffs and road construction of energy policy and air pollution.

As a rule, the "first best" solution would involve cutting back on those incentives which lead to a mismanagement of natural resources wherever this is possible. In certain cases this may be impossible as a result of conflicting goals, so that additional incentives have to be created. In the case of agricultural land management and agricultural policy, however, there is certainly more harmony than conflict between economic and ecological goals. It has often been demonstrated that by attempting for years to improve farmers' incomes with high farm prices the EEC agricultural policy has led to massive distortions in allocation. This has disturbed the economic balance as well as the ecological equilibrium (BALDOCK, 1985; v. MEYER, 1985).

XIV. INSTRUMENTS: THE IMPLEMENTATION OF NEW INCENTIVES

If one were to decide not to correct the inappropriate incentives and to rely solely on new instruments for land management, this would merely be a "second best" solution. If one wishes to buy or lease nature conservation areas or to compensate for the non-use of fertilizers and pesticides by paying aid for loss of yield, then this would be substantially more expensive under a protectionist price policy than under a more liberal one.

The argument for giving priority to a reorientation of agricultural policy should, however, not be misunderstood. Deciding to do entirely without additional new incentives would not bring about better land management in the future. That is to say that one cannot expect the misdirected structural change which has been induced by distorted price relations to be nullified by a short-term correction in price policy. Many of the changes which have been made are probably irreversible even in the medium term. Moreover, because of the technological changes which have occured, a reversion to market prices would not be accompanied by a reversion to the old ecologically desirable practices.

Furthermore, on equity grounds the use of supporting measures cannot be dispensed with. The political acceptance of the reforms required will not be attainable without providing social security. The legitimacy for additional instruments can thus be found (v. MEYER, 1987 a, b) not only for social aid but also for environmental (recreational and ecological) services that have not been remunerated up to now.

It would appear to me that the concept of "cross-compliance" which is at present being propagated in the United States could be a sensible approach for a longer transition phase. An attempt is being made here to make agricultural policy support dependent on compliance with environmental conditions (ERVIN/HEFFERNAN/GREEN, 1984; CLARK/RAITT, 1986; BATIE/SAPPINGTON, 1986).

The first moderate signs of the creation of economic incentives for an improved form of land management may be detected in some of the EEC's new measures on agricultural structures policy, such as the "reconversion" of land and "extensification". But environmental policy aspects still do not have enough emphasis here (EEC COMMISSION 1985, 1986, 1987).

XV. INSTITUTIONS AND ECONOMIC INCENTIVES

In many cases one of the main hindrances to the use of economic incentives is an inadequate institutional framework. I would like to stress two aspects in particular here. The first is that the clear definition of property rights is a necessary requirement for effective land management. The second is that it is also important to arrive at an optimal distribution of administrative responsibilities for land management both by varying horizontal (different policies) as well as vertical (European, national, regional) aspects.

XVI. ESTABLISHMENT OF PROPERTY RIGHTS

A necessary prerequisite for the succesful use of economic incentives is the clear definition of the property rights of those involved. Frequently the discussion becomes polarised into the two simple alternatives of "private" and "public" property, with most economists having a preference for solutions based on the concept of "private" property. Often the wide range of structures and of assignment of entitlements is overlook (BROMLEY, 1978). My personal conjecture is that with regard to land management co-operative forms of land use, for example, which often have a very old tradition in agriculture, could be very effective (v. MEYER, 1987 a).

Effective land management may also require that a clear distinction be made between rights over land area and those over soil eco-system use.

With regard to agriculture and the environment in the Federal Republic of Germany the conditions are by no means clear. According to the Nature Conservation Act "orderly agriculture" is not an encroachment on nature. Since "orderly" usually means the present customary form, agriculture is conceded very far-reaching rights, so that the realisation of the goals of nature protection are only possible by means of costly subsidies, if at all (SRU, 1985).

The argument in Germany about ground water protection is exemplary here. Whilst, on the one hand, there is a demand for high taxes of 100 per cent on nitrogen fertilizers (SRU, 1985; WEINSCHENK/GEBHARD, 1985), on the other hand, since the new amendment to the Water Management Act it is now necessary to pay compensation to farmers in water catchment zones for yield losses caused by the limitations on the "orderly" use of fertilizers and pesticides (BONUS, 1986; v. MEYER, 1987, c).

XVII. DISTRIBUTION OF ADMINISTRATIVE RESPONSIBILITIES

The optimal distribution of administrative responsibilities is also important if economic incentives are to function properly. A difference ought to be made between aspects of horizontal and vertical co-operation and co-ordination.

Horizontal differentiation involves the relationship between resource management and agricultural policy, regional policy, transport policy and energy policy, etc. A lot is to be said here for integrating the defining of instruments, a clear separation of aims would seem to be more appropriate. In practice, however - as the example of "cross-compliance" shows -there would be frequent co-operation and often policies could be designed so that conflicts would at least be kept to a minimum.

The example of the different administrative structures in the eleven German Federal States (Länder) also shows that there are positive as well as negative sides to a separation of environmntal and agricultural ministries and administrative bodies.

The vertical distribution of responsibilities involves the question of European, national and regional responsibilities. There is no doubt that a common agricultural market and price or a common definition of - not necessarily uniform - environmental standards for the EEC is not only sensible but necessary. In the light of the large regional differences in the natural, economic, social and legal conditions and structures, totally different requirements and possibilities for environmental action are required.

Efficient resource management therefore presupposes regionally differentiated action. As experience has widely shown that this cannot be realised by direct control from central authorities, it would appear more sensible to create organisational structures in which increased regional autonomy can contribute to the improvement of land management.

REFERENCES

1. Baldock, D. (1985): The CAP Price Policy and the Environment - An Exploratory Essay, in: Baldock, D. and Conder, D. (eds.) Can the CAP fit the Environment?, London, pp. 55-74.

2. Batie, S.S. and Sappington, A.G. (1986): Cross-Compliance- As a Soil Conservation Strategy: A Case Study, in: American Journal of Agricultural Economics. Vol. 68, pp. 880.885.

3. Bauwens, A. and Douw, L. (1986): Rural Development. A Minor Problem in the Netherlands? in: European Review of Agricultural Economics, Vol. 13-3, pp. 343-366.

4. Bmelf - Bundesministerium für Ernährung, Landwirtschaft und Forsten (1986a): Statistisches Jahrbuch 1986, Münster-Hiltrup.

5. Bmelf (1986b): Waldschäden in der Bundesrepublik Deutschland - Ergebnisse der Waldschadenserhebung 1986, Angewandte Wissenschaft, H.334, Münster-Hiltrup.

6. BMI - Der Bundesminister des Inneren (ed.) (1985): Bodenschutz - Konzeption der Bundesregierung, BT - Drucksache 10/2977, Bonn 7.3.1985.

7. Bonus, H. (1986): Eine Lanze für den "Wasserpfenning" - Wieder die Vulgärform des Verursacherprinzips, in: Wirtschaftsdienst, 66. Jg., H. 9, pp. 451-455.

8. Bowers, J.K. (1987): Set-aside and other Stories, Paper presented at the IEEP/CPRE Seminar, London, 28 January 1987.

9. Bromley, D.W. (1978): Property Rules, Liability Rules and Environmental Economics, in: Journal of Economics Issues, Vol. XII, No. 1, pp. 43-60.

10. Cheshire, P. (1985): The Environmental Implications of European Agricultural Support Policies. In: Baldock, D. and Conder, D. (eds.): Can the CAP fit the Environment? London, pp. 9-18.

11. Clark, R.T. and Raitt, D.D. (1986): Cross-Compliance for Erosion Control: Anticipating Efficiency and Distributive Impacts: Comment, in: American Journal of Agricultural Economics Vol. 68, pp. 1013-1015.

12. Countryside Commission (1987): New Opportunities for the Countryside, CCP 224, Cheltenham.

13.   EEC - Commission (1985):
       -- Perspectives for the Common Agricultural Policy (COM (85)
       333 final), Brussels, 15.7.1985.
       -- A future for Community Agriculture (COM (85) 750 final), Brussels,
       18.12.1986.

14.   EEC - Commission (1987):
       -- Proposal for a Council Decision Establishing a Community System of
       Aids to Agricultural Income (COM (87) 166 final), Brussels 22.4.1987.

15.   EEC - Commission (1986):
       -- Proposal for a Council Decision Amending Regulation Nos. 797/85 on
       Agricultural Structures (COM (86) 199 final/2), Brussels, 31.7.1986.

16.   Ervin, D.E., Heffernan, W.D. and Green, G.P. (1984): Cross-Compliance
       for Erosion Control: Anticipating Efficiency and Distributive Impacts,
       in: American Journal of Agricultural Economics, Vol. 66, pp. 273-278.

17.   De Haen, H. (1985): Interdependence of Prices, Production Intensity
       and Environmental Damage from Agricultural Production, in: Zeitschrift
       für Umweltpolitik, Vol. 8, H. 3, pp. 199-219.

18.   Hodge, I. (1986): The CAP and the Soil. A Preliminary Analysis, in:
       EAAE (ed.): 11th Seminar on Multipurpose Agriculture and Foresty,
       Padou 1986, pp. 357-370.

19.   House of Lords - Select Committee on the European Communities (1984):
       Agriculture and the Environment, House of Lords Paper 267,
       Session 1983-1984, 20th report, HMSO, London.

20.   IMAB - Interministerielle Arbeitsgruppe Bodenschutz (1984):
       Flächennutzung und Bodenschutz, Abschlussbericht der
       Unterarbeitsgruppe IV "Flächennutzung", vom 20.2.1984.

21.   Lee, J. (1986): The Impact of Technology on the Alternative Uses for
       Land, Fast-Occasional Papers, EEC-Commission, Brussels.

22.   V. Meyer, H. (1983): Wirkungslose Umweltpolitik - Umweltwirksame
       Agrarpolitik. Überlegungen zum Verhältnis beider Politiken in der
       Europäischen Gemeinschaft, in: Zeitschrift für Umweltpolitik, 6. Jg.,
       H. 4, pp. 363-387.

23.   V. Meyer, H. (1985): CAP and CEP - Discord or Harmony?
       - Environnemental Implications of Agricultural Price Policy and Reform
       Proposals: A German View, in Baldock, D., and Conder, D., (eds.): Can
       the CAP fit the Environment?, London, pp. 33-38.

24.   V. Meyer, H. (1987 a): Verteilungswirkungen einer umweltpolitisch
       motivierten Reform der Agrarpolitik, in: Schriften der Gesellschaft
       für Wirtschafts - und Sozialwissenschaften des Landbaues e.V., Bd. 23,
       Münster-Hiltrup (forthcoming).

25. V. Meyer, H. (1983): Wirkunglose Umweltpolitik - Umweltwirksame Agrarpolitik. überlegungen zum Verhältnis beider Politiken in der Europäischen Gemeinschaft, in: Zeitschrift für Umweltpolitik, 6. Jg., H. 4, pp. 363-387.

26. V. Meyer, H. (1987 b) Agricultural Income Policy - Why and How?, in: Alternative Support Measures for Agriculture, European Institute of Public Administration, Maastricht.

27. V. Meyer, H. 1987 c) Ansätze zu einer Agrar-Umweltpolitik in ökonomischer Theorie und politischer Praxis, in: V. URFF, W. and V. Meyer, H. (eds.): Landwirtschaft, Umwelt und ländlicher Raum - Herausforderungen an Europa, Baden-Baden.

28. OECD (1986) Water Pollution by Fertilizers and Pesticides, Paris.

29. Potter, C.A. (1986): Environmental Protection and Agricultural Adjustment: Lessons from the American Experience. Set Aside Working Paper No. 1, Wye College.

30. Priebe, H. (1982): Leben in der Stadt oder auf dem Land - Mehr Lebensqualität durch sinnvolle Raumgestaltung, Düsseldorf, Wien.

31. SRU - Der Rat Von Sachverständigen Für Umweltfragen (1985): Sondergutachten "Umweltprobleme der Landwirtschaft". BT-Drucksache 10/3613, Bonn, 3.7.1985. English Summary, Wiesbaden, March 1985.

32. StJB - Statistisches Jahrbuch für die Bundesrepublik Deutschland (1985): Stuttgart and Mainz.

33. Sukopp, H. (1981): Veränderungen von Flora and Vegetation. In: Berichte über Landwirtschaft, N.F., S.H. 197, pp. 255-264.

34. USDA - U.S. Department of Agriculture Soil Conservation Service (1982): A National Program for Soil and Water Conservation: Final Program Report and Environmental Impact Statement, Washington, D.C.

35. Weinschenk, G. and Gebhard, H.J. (1985): Möglichkeiten and Grenzen einer ökologisch begründeten Begrenzung der Intensität der Agrarproduktion, in: Rat Sachverständigen Umweltfragen (Hrsg.): Materialien zur Umweltforschung, Bd.11, Stuttgart, Mainz.

Chapter 10

APPLYING ECONOMIC RATIONALITY TO PROBLEMS OF RURAL LAND MANAGEMENT:

A EUROPEAN PERSPECTIVE

Dr. Paul Cheshire

I. INTRODUCTION

The purpose of this contribution is not to challenge the conclusions of Southgate (Chapter 8) but to argue, like von Meyer (Chapter 9), that there are additional aspects to the contribution of economic incentives to land management; and that in Europe these additional aspects -- relating to the economics of the rural environment and of conservation -- are considerably more critical than are issues relating to soil conservation. Because of climate, terrain and soil types, soil conservation has been a central issue in the United States for most of this century. To this European observer it appears that cultural, ideological and political factors have also been influential in bringing and keeping soil conservation to the fore there.

The critical questions in rural land mangement in Europe relate, however, to a much wider conception of the "rural environment" than the soil's mechanical ability to produce crops or the direct effects of agricultural pollutants on water supply quality. This is not to deny that in parts of Europe soil conservation is an issue of growing significance, particularly in the South and on some of the marginal land that high agricultural support prices have brought into intensive cultivation; in Sicily, for example, large tracts of rough upland grazing pasture and olive trees have recently been cleared and put to arable land. Soil erosion will inevitably follow as it does on the steep slopes of rough pasture in England and Wales that have been ploughed. Nor is it to deny that in the intensively farmed areas of Europe direct agricultural pollution of water supplies is a problem. Levels of nitrates in drinking water are very substantially above those recommended by the World Health Organisation or permitted by the European Commission in significant areas of Southern and Eastern England. There are problems of agricultural pollutants contaminating water supplies elsewhere in Europe also, quite apart from widespread biological degradation of waterways.

There is, however, a further dimension to the nature of problems of rural land use in Europe that hardly exists in the United States (or Canada or Australasia). By European standards the pressure of population on land in the United States is even now very low. The availability of rural land for recreation and wildlife has been less of an issue and the general solution has been for specialised uses to develop. Land is farmed and considerations of access, amenity or wildlife given virtually no weight; or it is dedicated to some widely-defined conservation use as a state or national park or perhaps as public forest. Given high per capita non-farm incomes, extensive land area and relatively less supported food prices, great tracts of less fertile non-urban land are simply not farmed commercially. The most obvious cultural concept of countryside in the United States is "wilderness".

II. FOOD OR RURAL AMENITY

The idea of "wilderness" areas -- or of deliberately reducing the agricultural margin and re-creating wilderness areas, tends to be greeted with strong hostility in Europe. The cultural ideal in prosperous Europe (and there appears to be a high income elasticity of demand for this cultural ideal) is much closer to "countryside". By this is meant a pastoral landscape of multiple uses; agricultural, recreational, touristic, sporting, wildlife habitat and water supply. In Germany, this cultural ideal is probably epitomised by the countryside of Alpine pastures. The Italians, French and perhaps most strongly of all, the British have their own versions of it. It is a potent image that can be used for very diverse purposes. It can be used to draw foreign tourists; it has been used by farmers' organisations to woo consumers and taxpayers as farmers lobby for more support; it has been used by advertisers to sell butter substitutes. Like food, non-food and water supply products of countryside can be classed as a composite good -- rural amenity. Rural amenity is produced as a joint product with agricultural and forestry use of rural land but its production is, certainly at the margin, non-complementary.

The supply of land relative to population is not the only factor in explaining these differences between Europe and North America -- historical patterns of land settlement and major structural differences in agriculture are also significant. Nor are the differences so absolute as this characterisation suggests. They amount, nevertheless, to a difference in kind. It is possible, also, that in Europe there exists another non-food and water demand for rural land -- 'social countryside'; that is, a demand by society as a whole that rural areas should be occupied on a full-time basis by people who earn their livings from rural (as distinct from ex-urban) pursuits. This seems to be an important element in fears expressed in France of 'desertification'; this does not mean the creation of physical deserts but the abandonment of the countryside. But, if such a demand exists, it is far less concrete than that for rural amenity.

III. PROBLEMS OF MARKET FAILURE AND INTERVENTION

The composite good rural amenity is not only a substitute, in production, for food, it is a quasi-public good for which no effective markets presently exist. Southgate rightly points out the classic problems with respect to public goods. Economic theory, however, deals in pure cases. A problem for policy is that in practice there exists a continuum. Some goods are private goods with rivalry in consumption and well-defined markets. Food is one such. Others are pure public goods. But many goods, including most in rural amenity, are neither one thing nor the other. There is an opportunity cost to other consumers of pure water but markets do not work all that well; markets are wholly ineffective in the case of informal access to the countryside but there is likely to be some small opportunity cost in consumption. Landscape may be a pure public good but recreational fishing may be marketed in some countries with particular definitions of property rights, while the definition of property rights may prevent that in other countries. There is always likely to be a problem of market failure, however, because of upstream agricultural (and, perhaps, industrial) pollution. In addition to these considerations there are likely to be two other sources of market failure in the provision of rural amenity. These are the 'option' and 'existence' demands discussed by Pearce (Chapter 1). Because of high transactions costs there is a strong expectation that markets will fail to give adequate weight to such demands. But in the case of rural amenity they demonstrably exist.

As a result of these problems of market failure there is, a priori, a presumption that in the absence of intervention, rural amenity will be under produced. There is, however, intervention on a grand scale. By far the most significant is agricultural support policies. Whether subsidies are on input or on output prices, they are highly regressive as income support measures since the larger the farmer is, the more support will be received. As a general income support mechanism for agriculture they are ineffective because, in the long run, they are capitalised into land values and, since land is a cost of production, returns are commensurately reduced. In the long run they are not, therefore, income supports but affect asset values or wealth. Traill (1982) estimates that in the United Kingdom a 1 per cent increase in output prices has been, in the long run, associated with a 10 per cent increase in land prices. Since price support is, however, an effective mechanism, both directly and indirectly (by promoting investment and research), in increasing food production and food production is a substitute for rural amenity, input and output price subsidies directly increase the underlying problem of market failure; the problem stemming from joint production where one product is sold through relatively effective markets and the other is effectively a public good.

The relationship between high prices (and subsidies) and environmental change is one that observers have long been casually aware of. Sir Robert Peel, whilst still a defender of Britain's Corn Laws, declared that amongst his major motives was:

> "[if the Corn Laws are repealed I would] view with regret cultivation receding from the hill-top which it has climbed under the influence of protection and from which it surveys with joy the progress of successful toil ... we should not forget ... that it is under the influence of protection to agriculture continued for two hundred years, that the fen has been drained, the wild heath reclaimed" (Sir R. Peel, 1839)

The underlying causal mechanisms are still present. What has changed is the relative social valuation of unploughed hill-tops compared to additional food production.

The mechanisms by means of which price support leads to a reduction of rural amenity were discussed in Cheshire (1986). A brief example may be taken to illustrate them. An important habitat and informal recreational resource is undrained pasture. We may take draining this as an example but a similar set of considerations applies to upland ploughing, treating meadows with fertilizers or herbicides, re-seeding permanent pasture, taking out hedges and small areas of scrub or ploughing footpaths or right up to hedge rows. Land is an input into agriculture but its price is determined by the profitability of farming. This is because the supply of land -- at least in Europe -- is highly inelastic, whilst the supply of capital tends to be highly elastic for all industries and the supply of labour to agriculture is highly elastic (which is why support policies have little effect on agricultural employees' wages relative to those of other employees).

Thus, in agriculture, price support on outputs changes relative factor prices and hence techniques. As was noted above, Traill (1982) estimates a 1 per cent increase in output prices generates a 10 per cent increase in land prices in the United Kingdom. The only way the effective supply of land can be increased as output and land prices rise is via intensification on existing land and extending the margin of cultivation (cultivation 'climbing the hill-top'). This, in turn, is associated with a substitution of capital for labour. Again relying on Traill's (1982) results for the United Kingdom, a 1 per cent increase in output prices generates an 0.4 per cent increase in capital stock and a 1.7 per cent reduction in labour inputs.

European support policies (before and after the CAP) have been typified by negligible incentives to farm amalgamation but by substantial price support and capital subsidies. The resulting substitution of capital for labour and intensification has been particularly associated with specialisation, therefore the economies of scale associated with more capital intensive methods of production can only be exploited, given quasi-fixed farm sizes, by reducing the number of farm enterprises. This has been re-inforced by the elimination of price fluctuations produced by price support. This has meant that the insurance advantages of multi-enterprise farms have been lost. In addition, the traditional economies of joint production, say of livestock and cereals, have also been lost, intensifying, for example, the problems of stubble burning and loss of habitat in specialist arable areas and the problems of manure disposal in intensive livestock areas.

Underdraining of meadows involves investment costs but produces higher yields of fodder per unit area coupled with higher inputs of capital and agro-chemicals since it is almost invariably accompanied by complementary investment in ploughing and re-seeding. It represents, therefore, the 'increase' in supply of land in the face of higher land prices via intensification and a move to a high input -- high output system. The initial investment has been indirectly induced by means of the price support (see above) and also, directly subsidised by investment incentives -- especially in Less Favoured Areas where actual payments for 'improvements' have frequently exceeded real costs. The process of intensification is associated with the substitution of capital for labour since, with quasi-fixed farm sizes, to exploit the economies of scale of, say, a milking parlour, it is necessary not only to specialise in dairy but to increase stocking rates. Another way of looking at it is that the more expensive land is and the higher output prices are, the greater the opportunity cost to the farmer of foregoing intensification. Thus high output prices, far from inducing a prosperous agriculture so that farmers can 'afford' to farm in a conservation minded way, provide powerful incentives to erode rural amenity.

The extent to which capital is substituted for labour varies not only with relative prices but also with technical conditions. Thus pigs and chickens are highly tolerant of intensive conditions and, despite low rates of price support, the rate of substitution of capital for labour and land has been exceptionally high. It must be remembered, however, that pig and chicken farmers are faced with similar changes in relative factor prices despite receiving little price support because they have to compete for land with farm enterprises that do receive such support. Sheep are intolerant of intensification so, despite high rates of support, intensification has tended to be less extreme (the problems of manure disposal are very slight or non-existent) and it has been via inputs of fertilizer, agro-chemicals, ploughing and re-seeding and higher stocking rates.

There are important variations in this process of intensification and the substitution of food output for that of rural amenity across Europe. In the United Kingdom, with its early bourgeois and industrial revolutions and different inheritance laws and institutional arrangements for land transactions, commercial agriculture developed very early. The rate of Parliamentary Enclosures between 1770 and 1830 was a direct function of wheat prices and a negative function of rates of interest and the amount of land not yet enclosed (see Chambers and Mingay, 1966). This compares closely with Baldock's (1984) finding that rates of under drainage are far higher in France than in Britain. In Britain the supply of non-drained land was far smaller.

Thus the rate at which the incentives of agricultural policy produce change and the totality of change itself, varies. The old established land owner (including the peasant farmer) can take part of the income available from price supports in a non-monetary form. Their decision making can react to the historic cost of land, not the current market value, so they can afford to be traditional and non-intensive or produce rural amenity for their own enjoyment. Thus in parts of Europe where such farmers are more dominant or land markets are highly imperfect, the rate and totality of change is much less and the supply of rural amenity more abundant. In Britain, with a highly developed land market and a predominantly commercial agriculture, intensification was far advanced by the late 1950s.

This explains why intensification frequently follows a change of ownership or management. The purchaser has paid the current market value and therefore quite suddenly agricultural systems become geared to current relative factor prices not historic ones. A similar process follows if the owner or inheritor uses the increased asset value of land either to convert the asset value to income or to borrow against to invest. This sometimes leads to the misguided conclusion that the loss of rural amenity is all a management problem which can be solved by education or goodwill. As the British agricultural economist, Nash (1965) put it in the context of already emerging national surpluses:

> "Since government has continued to pay them (farmers) handsomely for disregarding its advice it has inevitably created the impression either that it does not mean what it is saying or that it does not know what it is doing."

The result of economic incentives as applied to rural land use in Europe, then, is to significantly increase agricultural intensity and extend the margin of cultivation. This latter process is particularly associated with habitat destruction, restriction of access and loss of rural amenity.

As the conflict between food production and rural amenity has been increasingly understood (reflecting, as well as changing perceptions and values, underlying economic realities -- a diminishing supply of rural amenity, rising incomes and non-agricultural population, the induced substitution of commercial for traditional producers, and growing surpluses of food), so environmentally benign intervention has been brought onto the agenda, and even to a limited extent implemented. A major problem arises, however, if incentives are to be applied to influence agricultural techniques or the particular use of individual parcels of land with high 'amenity' value. That is why agricultural subsidies directly increase the cost of conservation payments. Calculations relating to the management agreements made possible by the Wildlife and Countryside Act in the United Kingdom show that typically up to 80 per cent of payments to farmers to refrain from intensification on environmentally sensitive sites, are subsidies to forego subsidies for producing food (Bowers and Cheshire 1983). Thus, if existing levels of intervention were reduced or made neutral in terms of their effects on food production compared to output of rural amenity, not only would the problems be significantly reduced but the costs of economic incentives to rectify the underlying imbalance in the output of food compared to rural amenity would be greatly reduced.

IV. THE ENVIRONMENTAL IMPACT OF LOWER LEVELS OF FARM SUPPORT

From this viewpoint, then, the priority with respect to economic incentives for conservation of rural land is the careful dismantling of the powerful economic incentives that presently exist to reduce rural amenity. Such action would slow the rate at which valuable habitats disappear, land is brought into intensive use, access restricted, or non-source pollutants increased. It may in general be expected to be associated with a move towards lower input -- lower output systems, although this may be a mean affect, possibly with intensification occurring in a few fertile areas. It is unlikely significantly to re-create the handmade countryside of the past or to re-establish traditional systems more complementary to rural amenity where these have disappeared.

Even without specific conservation incentives, however, it is likely that some structural changes resulting in increased joint output of food and rural amenity would occur. Historically lower land prices have encouraged an inflow of small scale farmers just as high output and land prices have been associated with an inflow of commercial, entrepreneur farmers, agricultural investment and intensification. There was an influx of 'improvers' drawn from the first generation of industrialists at the height of the Corn Law-protected prosperity of the first twenty years of the 19th century; the agricultural depression during the 1920s and 1930s was associated with an inflow of new, medium and small farmers and the break up of many large estates. Lower land prices reduce the price of entry (as well as making entry less attractive for pure profit seekers). Small farmers are intrinsically likely to be more productive of rural amenity because they tend to be less capital intensive. In addition, the objective functions of such entrants are more likely to include rural amenity. Some may be from traditional small farming families; others may be part time or have taken early retirement and gain direct welfare from their activities. Newby et al. (1977) provide evidence on the systematic variation in values and attitudes towards conservation of different types of farmers in the United Kingdom.

A second structural change that is occurring and would be likely to be accelerated by a significant reduction in overall support levels, is farm diversification. This is both into higher value added food products -- organic and/or better quality food and direct farm sales; and into the direct marketing of rural amenity or activities which depend upon rural amenity -- for example tourism and recreational activities. All these activities combined are presently quite small in proportion to conventional farming but command a large premium, apparently have high income elasticity of demand for their output and are either more compatible with rural amenity or directly complementary to it.

V. POLICIES FOR RURAL AMENITY

The case for economic incentives designed directly to enhance rural amenity -- primarily by increasing multiple land use and, by presumption, the joint output of food and rural amenity -- rests on conventional welfare economics (see the chapters by Pearce and Southgate in this volume). The efficacy of particular incentives has not been established by research. Some research has suggested that the demand for fertilizers is price inelastic, so taxes would not be effective (although, paradoxically it is quite likely their impact would be equitable -- large farmers paying proportionately the most and, as a group, being richer than consumers). This, like the research of the 1960s and early 1970s, that suggested demand for petrol was price inelastic, is usually based on purely-short term reactions and assumes output prices fixed. Allowing for long-term adjustments and reduced subsidies on output prices would suggest significantly higher price elasticities. The particular example cited by Southgate (Jacobs and Casler, 1979) suffers from the additional defect that the methodology of the study assumed rates of fertilizer application as given; the only response function estimated is that of variations in the cropping pattern. That the study found a low price elasticity was, therefore, inevitable. As well as incentives (and perhaps, regulation) to control fertilizer use there could be a case for incentives to recycle animal manure. One of the further peculiarities of agriculture is what might be described as the status quo. In the long run this is not exogenous but endogenous. Particularly given the structure of the industry there is a case for publicly funded research. This case is strongest, however, not where benefits can be internalised -- as, for example, with the application of agro-chemicals to plant strains bred to be high yielders with high rates of agro-chemical application -- but where the benefits cannot be easily internalised or cannot be internalised at all. There is thus a much stronger economic case for diverting incentives and public research efforts into recycling animal manure and plant breeding programmes designed to minimise the application of agro-chemicals than for the incentive and research structure that presently exists. Unfortunately, lobbyists are inevitably most active where benefits can be internalised and captured. There seems also to be a case for taxing energy consumption in agriculture at rates similar to those faced by other industrial users.

A whole range of amenity incentives are potentially defensible but again research is needed to show which would offer the greatest net gains. The evidence suggests that quite small adjustments in techniques of cultivation -- leaving unploughed and untreated headlands and hedge margins -- allowing small areas of scrub in inaccessible corners -- can dramatically affect the range of wildlife in the countryside as well as the supply of game, visual quality and access. The strongest case for incentives, however, may relate to access and agricultural and forestry practices complementary to access. This is because not only can no market for informal access exist but it may have some opportunity cost to owners of rural land, not only with respect to food output but also for some other elements of rural amenity; for example sporting activities and marketed recreational activity. In general the more intensive the agricultural system, the higher the opportunity cost of access to the land owner/manager; so again the arguments favour policy changes designed to foster low input-low output systems.

The purpose of this paper has not been to offer precise solutions but to raise a set of issues which are particularly relevant in the European context but which, as incomes rise and pressure on land increases, may be of increasing importance in other OECD countries; and to suggest a general pattern for approaches to the problem. This general pattern is to structure incentives -- including the elimination of existing incentives -- in ways which maximise net welfare from the multiple use of rural land. It is, emphatically, not to layer incentives in an unco-ordinated way. In the United Kingdom -- with separate government departments involved -- there is now a situation in which farmers in Environmentally Sensitive Areas are offered large incentives to intensify food production via the normal mechanisms of the CAP; then there are payments to forego those subsidies and farm in an environmentally benign way; and then yet further payments via a different scheme to produce no food and turn their land to intensive conifer plantations for timber. The priority is to rationalise the underlying system and the price signals which farmers receive, and so that they intrinsically favour rural amenity because markets by themselves will not and present policy certainly does not.

There is a further problem. That, as is mentioned elsewhere in this volume, is the problem of getting from here to there. It seems likely that a sudden dismantling of existing incentives apart from generating serious dislocation and financial instability might well have environmentally damaging results, particularly in the short term, in more fertile areas. That is why a careful dismantling may be necessary with incentives designed to enhance rural amenity gradually replacing those inducing intensification. There might also be a case for some form of short term assistance to farm owners who had borrowed heavily to invest.

There is, however, another danger. The announcement of set-aside programmes in the United States has been associated with an increase in sod busting on marginal land in order to qualify for extra set-aside payments. In the United Kingdom compensation payments to farmers to refrain from intensification on environmentally sensitive sites have been accompanied by 'threat' behaviour; an increase in declared intentions to intensify in order to get payments for not intensifying. Thus the lesson to be drawn from the present situation of failed intervention is not that invervention is wrong and cannot work; anymore than the lesson to be drawn from the fatal plunges of those who stuck bird feathers to their arms and tried to fly was that heavier than air flight was impossible. It is that economic and social processes are highly complicated and market forces can never be ignored even when the intention is to modify their results. Intervention must be carefully conceived with clear and consistent goals and must be planned and implemented in ways that take as full an account as possible of the ways in which economic agents react to incentives and to anticipated controls.

NOTES AND REFERENCES

1.   Contrary to received wisdom, the level of price support was higher in the United Kingdom than in any of the other countries in the EEC 10 prior to the establishment of CAP. McCrone (1961) estimated United Kingdom support to be 50 per cent higher than that in the next most protective country -- France.

2.   This, however, is extremely difficult to achieve because of the need to avoid the capitalisation of support into land values if intensification is to be avoided as a response to rising land prices. The most promising method perhaps is to offer means tested income supplements in which farmers have no property rights. They could perhaps be administered on a discretionary basis by an agency whose brief was countryside policy not food production. Thus income supplements might be conditional on management agreements designed to enhance rural amenity.

3.   Baldock, D. et al. (1984), Wetland Drainage in Europe: the effects of agricultural policy in four countries, Institute for European Environmental Policy, International Institute for Environment and Development, London.

4.   Bowers, J. K. and Cheshire, P. C. (1983), Agriculture, the Countryside and Land Use, Methuen, London.

5.   Chambers, J.D. and Mingay, G.E., The Agricultural Revolution 1750-1880, London, Batsford.

6.   Cheshire, P.C. (1986), 'The Environmental Implications of European Agricultural Support Policies', in Baldock, D., and Conder, D. (eds.), Can the CAP Fit the Environment?, Council for the Protection of Rural England and Institute for European Environmental Policy, London.

7.   Jacobs, J.L. and Casler, G.L. (1979), 'Internalising Externalities of Phosphorous Discharges from Crop Production to Surface Water: Effluent Taxes versus Uniform Reductions', American Journal of Agricultural Economics 61, 2, pp 309-312.

8.   McCrone, G. (1961), The Economics of Subsidising Agriculture, George Allen and Unwin.

9.   Nash, E.F. (1965), 'A Policy for Agriculture' in McCrone, G. and Attwood, A.E. (eds), Agricultural Policy in Britain: Selected Papers by E.F. Nash, Cardiff, University of Wales Press.

10.   Newby, H. et al (1977), 'Farmers' Attitudes to Conservation', Countryside Recreation Review, Vol 1. Peel, Sir R. (1839) On Mr. Villier's Motion on the Corn Laws, John Murray, London.

11.   Traill, B. (1982), 'The Effects of Price Support Policies in the United Kingdom', Journal of Agricultural Economics, Vol XXXIII, No. 3, pp 369-385.

Chapter 11

CONCLUDING COMMENTS

Dr. Daniel W. Bromley
and
Prof. David W. Pearce

These concluding comments on the discussions at the workshop are grouped in four sections:

1) General considerations for improved management of natural resources;

2) The importance of varying perceptions of natural resource problems;

3) Issues and resource problems identified by the workshop participants; and

4) Policy opportunities.

I. CONSIDERATIONS FOR RESOURCE MANAGEMENT

Efficient natural resource management is a 'sine qua non' for efficient and cost-effective environmental management for both present and future generations.

Inefficient use results in excessive depletion of exhaustible resources, e.g. soil erosion, and incurs unwarranted costs of regenerating renewable resources, e.g. forests. Inefficient use also leads to more pollution, e.g. cheap water often generates more water pollution or aggravates salinisation of the soil. In some cases these damages are irreversible.

Good management or husbandry of natural resources on the other hand can eliminate or reduce these effects, often with savings both in terms of natural resources and of reduced pollution damage.

Natural resources management should reflect in a balanced way the multiple uses to which these resources can be applied.

Many of the problems addressed at the workshop arise from the multiple output and use of renewable natural resources and the competing demands which exist for these alternative uses.

Water might be used as drinking water, for industrial use or for irrigation; it also provides transportation and recreational facilities. Land provides a multitude of beneficial services to a society -- as location, as amenity, and as productive soil. A plot of forest land may be used to produce wood, to protect soil, to provide wildlife habitat and recreation, and as a symbol of national heritage. It follows that different individuals will have different ideas about the proper management of that resource. These overlapping uses and objectives complicate the resource management task.

Countries should aim at managing renewable resources on a sustainable basis:

-- that will assist long-term economic growth;

-- that will give due regard for environmental quality and for quality of human life.

There is, as yet, no concensus on the proper definition of sustainability. Yet in the environmental context sustainability of natural resources is important when we recognise that these resources can be irreversibly damaged from over-use or from improper management. A groundwater aquifer is a renewable resource, but if subject to repeated excessive withdrawals it will lose its ability for recharge. Likewise, once an aquifer is contaminated with toxic compounds it may be centuries before it can be cleansed. A government is the principal social entity concerned with assuring the continuation and longevity of the society of which it is a part. In that sense, government ought to take an interest in resource sustainability on behalf of those citizens not yet born. Governments should seek to avoid those situations in which large irreversible losses result from policy formulation. Natural resource management can be understood to be part of this commitment to the future.

Natural resource management policy should be concerned with a set of policy objectives and a mix of policy instruments selected to achieve efficiently the stated objectives.

It is for the government to set down broad policy objectives and decide on the mix of policy instruments. The policy objectives are likely to be the same whether natural resources are privately owned and managed by the markets or are in public hands.

Each country has its own preferred system of allocating natural resources to the public and private sector. Allocation of ownership is still a source of problems in itself, but while the allocation between various types of ownership is important for efficient management, the clear definition of objectives and the proper mix of instruments to achieve them is more important.

It is obvious that the private sector and government have different imperatives, different time horizons, and different performance criteria.

Analysis that holds government performance up to the standards of the profit-maximising entrepreneur may not be entirely valid, and also may often be misleading. Similarly, to expect the private sector voluntarily to consider offsite costs and the interests of future generations is unrealistic. Both sectors are good at their own tasks and appropriate policy instruments and evaluative criteria pertinent to each are needed for effective overall management.

Clear definition of property rights, irrespective of private or public ownership, can facilitate good management. Furthermore these property rights, in particular the right to use, should be designed to allow the transfer of the resource to alternative, more beneficial uses, both in economic and environmental terms.

Property rights are in constant evolution. In some OECD countries, for example, water rights are being radically rewritten. There is also in most countries a need for a clearer definition of groundwater rights. The right to use land for different purposes is also constantly changing either due to zoning or to environmental needs.

In particular in the field of water, major misallocations with substantial environmental damage could in some cases be avoided if property rights, or rights to use, could be transferred, for example, between regions.

Pricing of natural resources is one of the major instruments of natural resource policies. Prices are, depending on the nature of the product, set by the market and by the public sector. Governments in general should avoid subsidising resource prices, because subsidises usually lead to overuse of the resource and more environmental damage than otherwise.

Underpricing of water, for example, in agriculture, had led not only to excessive use, but also to environmental degradation. Subsidised timber production can result in over-planting especially of monocultural species, damage to water-sheds or destruction of wilderness areas. Subsidisation of land use, e.g. agriculture, could also lead to various forms of over-exploitation or to the reduction of wetlands and other areas of special ecological value.

There would seem to be strong arguments for a re-appraisal of existing pricing policies together with the various incentives provided by governments to improve natural resources management. Such a re-appraisal could benefit from a joint effort by interested OECD Member governments.

The assessment of resource management problems and of the policy outcomes is the task and responsibility of governments.

This assessment will require an evaluation of the physical as well as the economic impacts. Some of these physical impacts can be expressed in monetary terms, while others, as of yet, cannot. There should be no presumption that monetised impacts are more pertinent or "real" than non-monetised impacts; monetisation should simply serve to elaborate information used in the decision process.

The assessment process should distinguish between those resource management problems that are reversible from those that are not reversible. The assessment should also include information about the urgency with which each problem ought to be dealt. That is, some problems imply an added financial burden on a sector of the economy, others may manifest themselves as irritants or nuisances, while yet others will pose long-term implications for human health and for the ecological integrity of a country.

Governments are responsible for formulating their objectives and implementing their policies in a consistent and integrated manner.

Consistent and integrated policy formulation theoretically emerges from the political process reflecting citizens concerns for policies. Inevitably, there will be some conflicts between those aiming for short-run rapid economic growth and a certain type of income distribution, and specific environmental objectives. Ideally, the political process is designed to minimise these contradictions and make policy objectives mutually compatible.

Efficient implementation of policy is ensured through the proper integration of policy instruments and in particular of various incentives. This requires also considerable integration and co-operation among the various governmental agencies involved.

From the environmental point of view, it can be argued that such integration could assign a leading role to environmental agencies in areas where health issues are predominant, and a more balanced distribution of responsibilities where purely efficiency or amenity aspects are involved.

The costs and benefits of resources management policies, both environmental and economic, should be evaluated prior to implementation and governments should be mindful of the incidence of the costs.

While economic efficiency is one of the main concerns in policy formulation the incidence of social costs accompanying these policies is also of relevance. These costs shift in time, and their impact is spread both across groups and interests now, and across generations. As a minimum, information is needed as to how these cost shifts are spread before a political consensus can be reached on their desirability.

II. VARYING PERCEPTIONS OF THE RESOURCE MANAGEMENT PROBLEM

The opportunities to resolve resource management problems will vary among Member countries. During the Workshop several factors were identified that will lead to different perceptions of resource problems, to different policies being pursued, and to different judgments of the ideal policy strategy. These differences are listed below.

-- Varying status quo systems of resource rights tend to influence perceptions of feasible policy. At present, wholesale redefinitions of property rights are of policy relevance only in some countries but limited changes, in particular in the right to use resources, could significantly improve policy making.

-- Varying physical and topographical environments may militate against policy standardisation across Member countries. The potential for redefining property rights in water, for example, is not the same in countries with limited, well-defined source points as in countries where sources, interactions and rates of water re-use are many and varied. What is possible for Australia, say, may not be possible for the Netherlands.

-- The availability of 'frontier territory' in some countries offers greater scope for land use based on wilderness concepts of conservation. Smaller, more densely populated countries are more likely to pursue habitat preservation in terms of closely managed recreational use. Europe, for example, has few remaining natural wetlands, and conservation of what remains is very likely to be along the lines of integrative recreational use rather than wilderness 'set-aside'.

-- Allied to the population density issue, environments that are enjoyed rather than "consumed" have grown in importance. We hazard the view that low density frontier economies still tend to be subject to the perception of environment as a usable commodity since use rates are often well below the carrying capacity of the environment. High density 'spaceship' economies attach higher and higher marginal values to the ever dwindling stock of natural environments, perhaps as psychological contrast to, and relief from, urban stress.

-- The perception of the very nature of the countryside is strongly influenced by the character of environmental threats. While soil erosion unquestionably exists in Europe, Europeans do not see it as a major threat to agriculture and hence to the countryside. Their perception is heavily influenced by urban sprawl: countryside is a matter of quality and quantity.

For these reasons, and no doubt others, only certain aspects of rational resource management, e.g. pricing policies, can be standardised across Member countries, and the potential for policy-imitation accordingly lessened. As Section I stressed, policy mixes are likely to be as varied as the natural and built environments are varied. Nonetheless, comparative experience is suggestive and the study of such experience must be a priority within OECD.

III. CURRENT RESOURCE MANAGEMENT ISSUES AND PROBLEMS

1. Water Resources

Water resources are managed in the OECD countries under a variety of ownership arrangements. The prevailing view is that water is a gift of nature and as such ought to be publicly owned and controlled. In other settings water rights are held by individuals under several forms of legal tenure. The expression of individual rights over water is usually, though not always, associated with the taking of water out of its natural location for uses that may or may not be consumptive in nature. Agriculture is, for the most part, regarded as a consumptive use of water. Industrial and domestic uses are generally regarded as non-consumptive, though of course both uses can seriously impair the quality of water and thereby reduce its useful quantity.

Thus we find countries in which:

1) Water "belongs to" the government and is managed by the government in the interest of current and future citizens; and

2) Water rights are bestowed to individuals as long as their use of it does not conflict with an overriding "public interest".

Even where private rights to water exist we will find some "reserved" or "residual" interest being retained by the state. In those countries where water is the exclusive domain of the state the management problem is to create a constellation of policy instruments that will result in a pattern of water uses meeting several general criteria. We propose that such criteria (policy objectives) will include the sustainable use of water resources, the promotion of economic growth, and the achievement of these objectives at the lowest possible administrative costs. The policy instruments will probably be a combination of regulations and of price-based incentives and sanctions.

While economics would suggest that these two different social attitudes about ownership and control of water resources will result in two different paths of economic growth, we cannot label one as socially preferable to the other since judgments about social optimality must be based on each country's policy goals and objectives. Each country will therefore choose its own policy objectives for water resource management. OECD could however make any important contribution with respect to the specification of those objectives, and with respect to the appropriate policy instruments.

OECD Member countries might wish to proceed jointly in specifying particular objectives, and in reaching a better understanding of the benefits and costs of alternative policy objectives.

Such specification of policy objectives, and the design of policy instruments, will need to recognise that ground water and surface water will present very different problems, and very different policy opportunities. Water resource management has both a quantitative and a qualitative dimension -- and they are obviously related; when water quality deteriorates to the point that it is unusable then its diminution in quality is a diminution in quantity.

The policy problems in water resources will usually be seen to arise from several common causes. First is the absence of well-defined and consistent property rights in water. This means that there must be a clear link between the declared owner of water, and the managerial actions that are taken with respect to water resources. When water is the property of the state by official decree, yet there is no effective management and administrative structure in place to control its use (or misuse), then that clear linkage is missing and water management problems will certainly arise.

When water rights are vested in individuals the presumption is that markets will emerge to assure that water is allocated to its "highest and best use". But of course such perfect markets will not automatically arise and so the danger is that market-based allocations will not be driven by the proper social accounting (or shadow) prices. This is particularly serious when irreversible actions in water management hold implications for future generations who are not here to have their interests represented in market decisions.

Frederick (Chapter 2) identifies the substantial distortions arising from de facto policies of improper pricing of water resources. In turn, zero or near-zero pricing for in situ water is a direct, but not inevitable result of state ownership. But Frederick identifies a second problem, namely the multiple uses of water resources. Within the spectrum of uses, some have value (e.g. wildlife habitat) which is difficult to identify and measure, or which appear as mixes of private and public goods, greatly complicating the design of an efficient property rights system. Frederick shows how the opportunity to design such a system was lost in the United States. Market failure thus characterises property rights solutions based on individual ownership. The mere fact of repeated re-use, especially in population-dense countries, means that quality and quantity are materially affected as rates of use increase. Frederick also draws attention to the public goods aspects of water, as amenity for example, although a 'club good' is perhaps a more accurate description given the limitations to joint consumption set by overcrowding. The standard criticism of private ownership solutions is that the characteristics of public or club goods lead to non-revelation of true preferences due to the 'free rider' problem: if the good is supplied to any one consumer it will be supplied to all, and hence individuals have an incentive to understate their true demand. How serious under-revelation of preferences is in practice is questionable. Evidence from contingent valuation studies which seek to measure the free-rider distortion (or 'strategic bias') leads us to think that individuals do not, after all, distort preferences significantly in the presence of public goods.

Land Resources

Land resources and water resources are inextricably bound together and their separation here is artificial and for expositional purposes only. The essence of land use problems is found in differing attitudes about how particular prices of land should be used, and in the implications of current and possible future land uses for neighbours, or for future generations. Agriculture, as a dominant use of land, results in a rural landscape that some individuals may find unappealing. Hence land-use problems can concern aesthetic aspects as well as direct physical attributes. The multiple -- and conflicting -- interests in how land is used and how it is controlled imply that there will be multiple expressions of dissatisfaction with particular land-use outcomes.

Land provides a location for human activity, it has certain qualitative attributes with respect to some of those uses (soil quality and depth for agriculture, acidification for agriculture and forestry), and it is devoted to extensive uses such as forestry or agriculture, or to intensive uses such as urban activities. Policy will tend to be concerned with the actual uses to which land is put, the spatial proximity of those uses, or land as a receptacle for the depositions of other activities.

In those countries where population density is very low multiple uses of land are 'extensified' in that some land is exclusively agricultural, while other land is exclusively recreational or is classified as "wilderness". Where population density is very high those multiple uses are intensified in that they may occur on the same piece of land. The use of English agricultural land for both agriculture and recreation in the form of walking (rambling) is an example of this. Land-use policies to deal with conflicting multiple uses therefore need to be fashioned to reflect the particular context in which these conflicts arise.

Forest Resources

Forestry represents a particular form of land use in the OECD countries that is characterised by extreme variability in conditions of ownership and management. In some countries forest lands are primarily privately owned; in other countries forest lands are almost exclusively in the public sector. Finally, there are examples of a mixture of private and public forests.

For the most part, today's forests are the result of specific land-use policies that have perpetuated the existence of natural forests, or that have caused the reestablishment of former forests following harvesting. While the objectives of forest management will differ across countries, the one common element across virtually all forests is their multiple use. That is, forests provide not only a supply of wood for commercial and household uses, but they also provide habitat for wildlife, watershed protection, and human recreation opportunities.

The management problem in forestry is compounded by these multiple -- and often conflicting -- interests in how forests are to be used. The economically efficient management regime for commercial wood production will usually be quite different from the preferred use regime for the purpose of watershed protection, and this latter use may then conflict with the forest's value as a recreation site or as wildlife habitat.

In those settings where national control of forests has been in operation it may happen that policies do not change in response to new scientific knowledge about forest management. The familiar concern with fire prevention is now known to have fostered a particular species mix that is quite vulnerable to more damaging fires. Policies toward fire as a management tool (a policy instrument) will probably need to change to reflect this new knowledge.

It is also obvious that fiscal policies in the OECD countries will have an impact upon the vegetative cover of forest lands, as well as upon the timing of harvests. Such policies will often warrant co-ordination with other land-use (and forestry) objectives. As agricultural production in many countries continues to exceed domestic needs there is growing recognition that some agricultural land could and should be devoted to non-agricultural uses. Whether that land goes into forests or into other types of vegetation/use will become a major land policy issue in the future.

Moreover, it is not just an issue of forest or agriculture, but the type of forest and the type of agriculture. Much of the European debate about extending forestry is not about acreage so much as about finding incentives to diversify species to avoid monocultural pine plantations. This reflects high valuations for recreation and habitat over short-rotation wood production, as well as the aesthetics of 'square' forests. Again, it is no accident that the nature of the debate varies with population density and the availability of natural forest.

How far any tinkering with property rights can solve this kind of problem is very questionable, not least because a sizeable proportion of the benefits may well accrue to unrepresented populations -- notably future generations. The ideal, as Sedjo (Chapter 5) notes, is an all-embracing management authority which maximises the net social value of the forest's multiproduct function. But this implies an ability to 'wipe the slate clean' and to redefine property rights along the lines suggested by Patterson (Chapter 3) for water in Victoria, Australia.

If future generations have rights, however, it is not obvious how they are accounted for in the kind of revision taking place in Victoria. As noted previously, there are also formidable obstacles to such redefinitions in many countries. The potential is, however, extensive in many less-developed countries where land and resource rights are in desperate need of clarification, definition and registration. Morever, integrated land management based on a 'spatial unit of account' such as the watershed can be practised in these contexts. Elsewhere, the problems of unravelling the status quo property rights, with all the attendant conflicts, may be overwhelming and we must continue to expect adjustments and attentuation of rights at the margin as the pressure on land grows.

The most frequent causes of policy difficulties in land-use policy arise for several reasons. First, there is a failure to appreciate the extent of incompatible land uses in close proximity to one another, and a corresponding failure to reduce the implications of these incompatibilities. Second, there is often a failure to understand the multiple facets of land -- as location, as a natural resource in its own right, and as a psychological amenity. Finally, as indicated previously, there is a failure to link land use with water use, and to design policies that will reflect this critical interdependence.

IV. POLICY OPPORTUNITIES

The workshop identified several types of existing management practices and policies that may require correction if mismanagement of natural resources is to be prevented in the future. First, there is a need to pay closer attention to the specification of rights and duties in natural resources.

There may be instances in which resources under the control of the state may be more properly managed as individual private property. Equally, there are situations where private rights to resources (e.g. as in agricultural land) result in outcomes that are judged to be undesirable. We suggest that Member countries pay particular attention to the ownership structure of natural resources, and that policy design begin with this most essential variable.

As noted previously, the basic problem to be handled by any redefinition of property rights -- whether via 'privatisation', 'socialisation', or adjustments with existing modes of property rights -- is the multiple use of resources. In turn, these must be defined to include options on future use ('option values' -- see Pearce in Chapter 1), and the rights of future generations. The workshop also acknowledged the responsibility of current generations to safeguard and transmit the common heritage or 'patrimony'.

Second, the matter of compensation will invariably arise when policy changes are discussed. Those who have been using natural resources in a particular manner will usually claim that they have a "right" to behave in that way and if they are to change they will require compensation. There are two dangers with this notion. First it leaves the state open to an implicit environment of extortion in which resource users must, in effect, be bribed to cease doing something that is socially harmful. Second, it means that the size of the public sector budget -- or more specifically that portion of the budget that can be set aside for such payments -- becomes an effective constraint on how much policy change can be achieved.

There is a fine distinction between the exercise of state power to create something for the "public benefit" (in which case fairness would suggest that those who thereby lose should be compensated), and preventing someone from engaging in activities that will create "public harm" (in which case the "fairness" of compensation is not obvious). When farmers are constrained from polluting groundwater with nitrogen runoff is it to create a public benefit, or to prevent the creation of a public harm? If groundwater is regarded as initially having been "pure" then a prohibition on its deterioration clearly cannot be said to create a public benefit; on this tack no compensation for restrictions on agricultural practices seems justified. If, on the other hand, groundwater is initially regarded as "polluted" then restrictions on continued nitrification may appear to be "creating" a public benefit in which case newly constrained farmers will suggest that the beneficiaries should pay them to cease their fertilizing practices.

The "polluter pays" principle would suggest that no compensation is called for if a restriction is imposed since then the farmer (polluter) is paying through the reduced yields arising from a lessened application of fertilizers. The "beneficiary pays" principle would seem to suggest that those who like clean water should pay for this new good, and the farmers would be inclined to say that they should receive this payment as compensation from the beneficiaries.

In point of fact, the two principles coincide when we realise that farmers have been the "beneficiaries" of services rendered by water flow through their soil, and by the extra benefits in the form of enhanced yields arising from fertilizer applications. They were receiving services for which they were not being charged. If they were to pay the full social costs of their use of nitrogen fertilizers, a cost that includes the losses arising from the degradation of groundwater quality, then they would face a much higher price for nitrogen fertilizer, they would use less of it, and the groundwater problem would, in all likelihood, disappear.

Both principles (polluter pays and beneficiary pays) suggest that the farmer should not be compensated for restrictions on fertilizer applications, and indeed ought to pay for the harm created. Whether fertilizer use is reduced through "market-like" instruments (in the form of higher fertilizer prices) or through regulatory instruments (an upper limit on allowable fertilizer -- or nitrogen use) the farmer is paying for the deterioration in groundwater quality arising from his activities. Whether an additional charge in addition to this income loss is called for will require analysis in each particular setting.

Lastly, justifiable compensation requires that the initial configuration of prices and incomes should itself be devoid of significant distortions. The most conspicuous example of improper compensation mechanisms superimposed on to grossly distorted price signals is the (European) Common Agricultural Policy. Moreover, as Cheshire (Chapter 10) shows, the compensation has to be offered only because of the price policy. An initial price distortion thus gives rise to an environmental distortion which then has to be corrected by making the original distortion bigger still! Having encouraged farmers to over-produce through price and quantity guarantees, some governments now offer compensation to farmers to set aside land for environmental purposes. A vital aspect of policy reform is the rigorous and critical inspection of the forces giving rise to the distortionary use of natural resources.

Finally, the workshop devoted considerable attention to the matter of pricing of resource services. As indicated immediately above, there are "market-like" incentives and sanctions to modify resource use, and there are "regulatory" mechanisms that tend to concentrate directly on quantities of items (as opposed to their prices). Correct pricing is about selecting the proper shadow prices for natural resources that will properly reflect their marginal social (as opposed to private/individual) value. The determination of these shadow prices is a complex and time-consuming task. In some settings the shadow price will be set infinitely high in which case a "regulatory" instrument will be selected in place of a "market-like" instrument. Governments will have their own reasons for preferring one over the other.

WHERE TO OBTAIN OECD PUBLICATIONS
OÙ OBTENIR LES PUBLICATIONS DE L'OCDE

ARGENTINA - ARGENTINE
Carlos Hirsch S.R.L.,
Florida 165, 4° Piso,
(Galeria Guemes) 1333 Buenos Aires
Tel. 33.1787.2391 y 30.7122

AUSTRALIA - AUSTRALIE
D.A. Book (Aust.) Pty. Ltd.
11-13 Station Street (P.O. Box 163)
Mitcham, Vic. 3132   Tel. (03) 873 4411

AUSTRIA - AUTRICHE
OECD Publications and Information Centre,
4 Simrockstrasse,
5300 Bonn (Germany)   Tel. (0228) 21.60.45
Gerold & Co., Graben 31, Wien 1 Tel. 52.22.35

BELGIUM - BELGIQUE
Jean de Lannoy,
Avenue du Roi 202
B-1060 Bruxelles   Tel. (02) 538.51.69

CANADA
Renouf Publishing Company Ltd
1294 Algoma Road, Ottawa, Ont. K1B 3W8
Tel: (613) 741-4333
Stores:
61 rue Sparks St., Ottawa, Ont. K1P 5R1
Tel: (613) 238-8985
211 rue Yonge St., Toronto, Ont. M5B 1M4
Tel: (416) 363-3171
Federal Publications Inc.,
301-303 King St. W.,
Toronto, Ont. M5V 1J5   Tel. (416)581-1552
Les Éditions la Liberté inc.,
3020 Chemin Sainte-Foy,
Sainte-Foy, P.Q. G1X 3V6,   Tel. (418)658-3763

DENMARK - DANEMARK
Munksgaard Export and Subscription Service
35, Nørre Søgade, DK-1370 København K
Tel. +45.1.12.85.70

FINLAND - FINLANDE
Akateeminen Kirjakauppa,
Keskuskatu 1, 00100 Helsinki 10   Tel. 0.12141

FRANCE
OCDE/OECD
Mail Orders/Commandes par correspondance :
2, rue André-Pascal,
75775 Paris Cedex 16   Tel. (1) 45.24.82.00
Bookshop/Librairie : 33, rue Octave-Feuillet
75016 Paris
Tel. (1) 45.24.81.67 or/ou (1) 45.24.81.81
Librairie de l'Université,
12a, rue Nazareth,
13602 Aix-en-Provence   Tel. 42.26.18.08

GERMANY - ALLEMAGNE
OECD Publications and Information Centre,
4 Simrockstrasse,
5300 Bonn   Tel. (0228) 21.60.45

GREECE - GRÈCE
Librairie Kauffmann,
28, rue du Stade, 105 64 Athens Tel. 322.21.60

HONG KONG
Government Information Services,
Publications (Sales) Office,
Information Services Department
No. 1, Battery Path, Central

ICELAND - ISLANDE
Snæbjörn Jónsson & Co., h.f.,
Hafnarstræti 4 & 9,
P.O.B. 1131 – Reykjavik
Tel. 13133/14281/11936

INDIA - INDE
Oxford Book and Stationery Co.,
Scindia House, New Delhi 110001
Tel. 331.5896/5308
17 Park St., Calcutta 700016   Tel. 240832

INDONESIA - INDONÉSIE
Pdii-Lipi, P.O. Box 3065/JKT.Jakarta
Tel. 583467

IRELAND - IRLANDE
TDC Publishers - Library Suppliers,
12 North Frederick Street, Dublin 1
Tel. 744835-749677

ITALY - ITALIE
Libreria Commissionaria Sansoni,
Via Benedetto Fortini 120/10,
Casella Post. 552
50125 Firenze   Tel. 055/645415
Via Bartolini 29, 20155 Milano   Tel. 365083
La diffusione delle pubblicazioni OCSE viene
assicurata dalle principali librerie ed anche da :
Editrice e Libreria Herder,
Piazza Montecitorio 120, 00186 Roma
Tel. 6794628
Libreria Hœpli,
Via Hœpli 5, 20121 Milano   Tel. 865446
Libreria Scientifica
Dott. Lucio de Biasio "Aeiou"
Via Meravigli 16, 20123 Milano   Tel. 807679

JAPAN - JAPON
OECD Publications and Information Centre,
Landic Akasaka Bldg., 2-3-4 Akasaka,
Minato-ku, Tokyo 107   Tel. 586.2016

KOREA - CORÉE
Kyobo Book Centre Co. Ltd.
P.O.Box: Kwang Hwa Moon 1658,
Seoul   Tel. (REP) 730.78.91

LEBANON - LIBAN
Documenta Scientifica/Redico,
Edison Building, Bliss St.,
P.O.B. 5641, Beirut   Tel. 354429-344425

**MALAYSIA/SINGAPORE -
MALAISIE/SINGAPOUR**
University of Malaya Co-operative Bookshop
Ltd.,
7 Lrg 51A/227A, Petaling Jaya
Malaysia   Tel. 7565000/7565425
Information Publications Pte Ltd
Pei-Fu Industrial Building,
24 New Industrial Road No. 02-06
Singapore 1953   Tel. 2831786, 2831798

NETHERLANDS - PAYS-BAS
SDU Uitgeverij
Christoffel Plantijnstraat 2
Postbus 20014
2500 EA's-Gravenhage   Tel. 070-789911
Voor bestellingen: Tel. 070-789880

NEW ZEALAND - NOUVELLE-ZÉLANDE
Government Printing Office Bookshops:
Auckland: Retail Bookshop, 25 Rutland Stseet,
Mail Orders, 85 Beach Road
Private Bag C.P.O.
Hamilton: Retail: Ward Street,
Mail Orders, P.O. Box 857
Wellington: Retail, Mulgrave Street, (Head
Office)
Cubacade World Trade Centre,
Mail Orders, Private Bag
Christchurch: Retail, 159 Hereford Street,
Mail Orders, Private Bag
Dunedin: Retail, Princes Street,
Mail Orders, P.O. Box 1104

NORWAY - NORVÈGE
Narvesen Info Center – NIC,
Bertrand Narvesens vei 2,
P.O.B. 6125 Etterstad, 0602 Oslo 6
Tel. (02) 67.83.10, (02) 68.40.20

PAKISTAN
Mirza Book Agency
65 Shahrah Quaid-E-Azam, Lahore 3 Tel. 66839

PHILIPPINES
I.J. Sagun Enterprises, Inc.
P.O. Box 4322 CPO Manila
Tel. 695-1946, 922-9495

PORTUGAL
Livraria Portugal, Rua do Carmo 70-74,
1117 Lisboa Codex   Tel. 360582/3

**SINGAPORE/MALAYSIA -
SINGAPOUR/MALAISIE**
See "Malaysia/Singapor". Voir
« Malaisie/Singapour »

SPAIN - ESPAGNE
Mundi-Prensa Libros, S.A.,
Castelló 37, Apartado 1223, Madrid-28001
Tel. 431.33.99
Libreria Bosch, Ronda Universidad 11,
Barcelona 7   Tel. 317.53.08/317.53.58

SWEDEN - SUÈDE
AB CE Fritzes Kungl. Hovbokhandel,
Box 16356, S 103 27 STH,
Regeringsgatan 12,
DS Stockholm   Tel. (08) 23.89.00
Subscription Agency/Abonnements:
Wennergren-Williams AB,
Box 30004, S104 25 Stockholm Tel. (08)54.12.00

SWITZERLAND - SUISSE
OECD Publications and Information Centre,
4 Simrockstrasse,
5300 Bonn (Germany)   Tel. (0228) 21.60.45
Librairie Payot,
6 rue Grenus, 1211 Genève 11
Tel. (022) 31.89.50
Maditec S.A.
Ch. des Palettes 4
1020 – Renens/Lausanne   Tel. (021) 635.08.65
United Nations Bookshop/Librairie des Nations-
Unies
Palais des Nations, 1211 – Geneva 10
Tel. 022-34-60-11 (ext. 48 72)

TAIWAN - FORMOSE
Good Faith Worldwide Int'l Co., Ltd.
9th floor, No. 118, Sec.2, Chung Hsiao E. Road
Taipei   Tel. 391.7396/391.7397

THAILAND - THAILANDE
Suksit Siam Co., Ltd., 1715 Rama IV Rd.,
Samyam Bangkok 5   Tel. 2511630
INDEX Book Promotion & Service Ltd.
59/6 Soi Lang Suan, Ploenchit Road
Patjumamwan, Bangkok 10500
Tel. 250-1919, 252-1066

TURKEY - TURQUIE
Kültur Yayinlari Is-Türk Ltd. Sti.
Atatürk Bulvari No: 191/Kat. 21
Kavaklidere/Ankara   Tel. 25.07.60
Dolmabahce Cad. No: 29
Besiktas/Istanbul   Tel. 160.71.88

UNITED KINGDOM - ROYAUME-UNI
H.M. Stationery Office,
Postal orders only: (01)873-8483
P.O.B. 276, London SW8 5DT
Telephone orders: (01) 873-9090, or
Personal callers:
49 High Holborn, London WC1V 6HB
Branches at: Belfast, Birmingham,
Bristol, Edinburgh, Manchester

UNITED STATES - ÉTATS-UNIS
OECD Publications and Information Centre,
2001 L Street, N.W., Suite 700,
Washington, D.C. 20036 - 4095
Tel. (202) 785.6323

VENEZUELA
Libreria del Este,
Avda F. Miranda 52, Aptdo. 60337,
Edificio Galipan, Caracas 106
Tel. 951.17.05/951.23.07/951.12.97

YUGOSLAVIA - YOUGOSLAVIE
Jugoslovenska Knjiga, Knez Mihajlova 2,
P.O.B. 36, Beograd   Tel. 621.992

Orders and inquiries from countries where
Distributors have not yet been appointed should be
sent to:
OECD, Publications Service, 2, rue André-Pascal,
75775 PARIS CEDEX 16.

Les commandes provenant de pays où l'OCDE n'a
pas encore désigné de distributeur doivent être
adressées à :
OCDE, Service des Publications. 2, rue André-
Pascal, 75775 PARIS CEDEX 16.

72380-1-1989

OECD PUBLICATIONS, 2, rue André-Pascal, 75775 PARIS CEDEX 16 - No. 44623 1989
PRINTED IN FRANCE
(97 89 01 1) ISBN 92-64-13194-9